IET TRANSPORTATION SERIES 43

AI Techniques in EV Motor and Inverter Fault Detection and Diagnosis

Other volumes in this series:

Volume 1	**Clean Mobility and Intelligent Transport Systems** M. Fiorini and J.-C. Lin (Editors)
Volume 2	**Energy Systems for Electric and Hybrid Vehicles** K.T. Chau (Editor)
Volume 5	**Sliding Mode Control of Vehicle Dynamics** A. Ferrara (Editor)
Volume 6	**Low Carbon Mobility for Future Cities: Principles and applications** H. Dia (Editor)
Volume 7	**Evaluation of Intelligent Road Transportation Systems: Methods and results** M. Lu (Editor)
Volume 8	**Road Pricing: Technologies, economics and acceptability** J. Walker (Editor)
Volume 9	**Autonomous Decentralized Systems and Their Applications in Transport and Infrastructure** K. Mori (Editor)
Volume 11	**Navigation and Control of Autonomous Marine Vehicles** S. Sharma and B. Subudhi (Editors)
Volume 12	**EMC and Functional Safety of Automotive Electronics** K. Borgeest
Volume 15	**Cybersecurity in Transport Systems** M. Hawley
Volume 16	**ICT for Electric Vehicle Integration with the Smart Grid** N. Kishor and J. Fraile-Ardanuy (Editors)
Volume 17	**Smart Sensing for Traffic Monitoring** Nobuyuki Ozaki (Editor)
Volume 18	**Collection and Delivery of Traffic and Travel Information** P. Burton and A. Stevens (Editors)
Volume 20	**Shared Mobility and Automated Vehicles: Responding to socio-technical changes and pandemics** Ata Khan and Susan Shaheen
Volume 23	**Behavioural Modelling and Simulation of Bicycle Traffic** L. Huang
Volume 24	**Driving Simulators for the Evaluation of Human–Machine Interfaces in Assisted and Automated Vehicles** T. Ito and T. Hirose (Editors)
Volume 25	**Cooperative Intelligent Transport Systems: Towards high-level automated driving** M. Lu (Editor)
Volume 26	**Traffic Information and Control** Ruimin Li and Zhengbing He (Editors)
Volume 30	**ICT Solutions and Digitalisation in Ports and Shipping** M. Fiorini and N. Gupta
Volume 32	**Cable Based and Wireless Charging Systems for Electric Vehicles: Technology and control, management, and grid integration** R. Singh, S. Padmanaban, S. Dwivedi, M. Molinas and F. Blaabjerg (Editors)
Volume 34	**ITS for Freight Logistics** H. Kawashima (Editor)
Volume 36	**Vehicular Ad Hoc Networks and Emerging Technologies for Road Vehicle Automation** A.K. Tyagi and S. Malik
Volume 38	**The Electric Car** M.H. Westbrook
Volume 45	**Propulsion Systems for Hybrid Vehicles** J. Miller
Volume 79	**Vehicle-to-Grid: Linking electric vehicles to the smart grid** J. Lu and J. Hossain (Editors)

AI Techniques in EV Motor and Inverter Fault Detection and Diagnosis

Yihua Hu, Xiaotian Zhang and Wangjie Lang

The Institution of Engineering and Technology

Published by The Institution of Engineering and Technology, London, United Kingdom

The Institution of Engineering and Technology is registered as a Charity in England & Wales (no. 211014) and Scotland (no. SC038698).

First published 2023

The Institution of Engineering and Technology
Futures Place
Kings Way, Stevenage
Hertfordshire SG1 2UA, United Kingdom

www.theiet.org

British Library Cataloguing in Publication Data
A catalogue record for this product is available from the British Library

ISBN 978-1-83953-762-2 (hardback)
ISBN 978-1-83953-763-9 (PDF)

Typeset by MPS Limited

Cover Image: Songsak rohprasit/Moment via Getty Images

Contents

List of figures

List of tables

Nomenclature

a_i	coefficients associated with support vectors
b_{CM}	the Coffin–Manson exponent
c_s	center of the stator, rotor, and rotation
C	covariance matrix derived from X
D	discriminator
D	the constant latent vector size through all of transformer layers
D_1, D_4	diode 1, diode 4
D_S	source-domain dataset
D_T	the target-domain dataset
D_{ST}	a common sample
D_C	common samples
E	expectation operator
E	the patch embedding projection
f_s	the sampling rate
Δf	frequency resolution
$f(t)$	original signal
G	generator
i_A	load current in direction towards the load
i_d	current d-component measurement in dq-coordinates
i_q	current q-component measurement in dq-coordinates
I_c	the collector current
j	scale parameter in discrete wavelet transform (DWT)
$J(w)$	the training error at any instance
k	the number of segments after each signal cutting
$k(x, x_i)$	kernel function
L_s	the length of each vibration signal
m	the number of iterations
N	length of the sequence
N	number of trees in a random forest
N_f	the number of cycles to failure
p_i	the output of the SoftMax layer for class i

$p{\cdot}v{\cdot}$	Cauchy principal value
pm	permanent magnet temperature (in °C) measured with thermocouples and transmitted wirelessly via a thermography unit.
P_n	distribution of random noise
Q	matrix consisting of eigenvectors of matrix A
sgn	sign function
S_1, S_4	switch 1, switch 4
t_d	blanking time
T	sampling period
T_{ON}/T_{OFF}	the ratio of switch turn on time and turn off time
T	the range of temperature cycling
T_0	a reference range
u	left singular vector
u_S, u_T	the hidden layer feature invariants between the source domain and the target domain
$u(t)$	function undergoing Hilbert transform
u_d	voltage d-component measurement in dq-coordinates (in V)
u_q	voltage q-component measurement in dq-coordinates (in V)
U	$N{\times}N$ square matrix (left singular vectors)
V_{AO}, V_{BO}, V_{CO}	pole voltage at phase leg A, B, C
V_{dc}	DC bus voltage
$V_{GE(Th)}$	the gate-emitter threshold voltage
V_{GE}	the gate-emitter voltage
$V_{new}(t)$	down-sampled vibration signal
$V_s(t)$	the original collected vibration signal
V^T	$M{\times}M$ matrix (right singular vectors)
W_N^{nk}	complex exponential term in DFT computation
x_f	fake sample
x_r	real sample
x_j^i	the output of neuron j in layer i
x_j^{l+1}	the input data of layer l
$x_p \in R^{N\times(P^2 \cdot C)}$	a sequence of flattened 2D patches
$X(n)$	finite-length sequence
$X(k)$	discrete Fourier transform (DFT) coefficient
X	matrix of original data with n rows and m columns
γ	parameter in the Gaussian radial basis function kernel
y_i	the actual data value
$\hat{y}_i$	the predicted data value
z	latent code or random noise

z_i	the input of the SoftMax layer for class i
$\mathbf{\Phi}(x)$	transformation function from N-dimensional space to Q-dimensional function space
ω	angular frequency
τ	time-dependent offset in DWT
$\varphi(t)$	wavelet function
λ	eigenvalue
v	eigenvector
Σ	matrix with eigenvalues on the diagonal
σ	singular value
δ	the impulse function
θ_d	parameters of the discriminator
θ_g	Parameters of the generator

About the authors

Yihua Hu is a reader at the King's College London, UK. He was previously the head of the Electrical Engineering Group at the University of York. He is a fellow of the IET, holds a Royal Society Industry Fellowship, and is a member of the UK Young Academy. He has published 120 journal papers in IEEE Transactions journals with an *H*-index of 55 in Google Scholar. He is the author of 15 patents.

Xiaotian Zhang is currently pursuing his PhD in electrical and electronics engineering at the University of York, UK. He received his BSc degree in electrical engineering from Hohai University, Nanjing, China, in 2018 and received his MSc degree in electrical engineering from Kings College London, London, UK, in 2020. His research interests include AI-supported EV electric powertrain health monitoring, fault detection, and safety improvement.

Wangjie Lang is currently pursuing his PhD in electrical and electronics engineering at the University of York, UK. He received his BEng degree in electrical engineering from the University of Strathclyde, Glasgow, UK, and Lanzhou University of Technology, Lanzhou, China, in 2020. His research interests include electrical machine fault detection and diagnosis and EV powertrain break engineering-based AI techniques.

Abbreviations

AC-GAN	auxiliary classifier GAN
ACO-DDBN	ant colony optimization–based discriminative deep belief network
AE	autoencoder
ANFIS	adaptive neuro-fuzzy inference system
ANN	artificial neural network
APF	the auxiliary particle filter
BEMF	back-electromotive force
BI	Bayesian inference
CART	classification and regression tree
CDBN	convolutional deep belief network
CDT	complex decision tree
CL	convolutional layer
CMFE	composite multiscale fuzzy entropy
CNN	convolutional neural network
CRA	cumulative relative accuracy
CRB	cumulative random behavior
CRNN	convolutional recurrent neural network
CS	compressed sensing
CTE	coefficient of thermal expansion
CTGAN	conditional tabular generative adversarial network
CWRU	Case Western Reserve University
CWTS	continuous wavelet transform scalogram
DA	data augmentation
DAE	denoising autoencoder
DAFD	deep domain adaptation neural network
DBN	deep belief network
DC-GAN	deep convolutional GAN
DCTLN	deep convolutional transfer learning network
DDC	deep domain confusion
DE	dynamic eccentricity
DFT	discrete Fourier transform

EKF	extended Kalman filter
DL	deep learning
ELM	extreme learning machine
EMA	exponential moving average
EMD	empirical mode decomposition
ESD	electrostatic discharge
ESVM	ensemble support vector machines
FDD	fault detection and diagnosis
FFBPNN	the feedforward back propagation neural network
FFT	fast Fourier transform
FC	fully connected
FL	fuzzy logic
FLOPs	the floating point operations
FMM	fuzzy min–max
GA	genetic algorithm
GAN	generative adversarial network
GAN-RF	generative adversarial network-supported random forest
GAP	global average pooling
GELU	Gaussian error linear unit
G-LS-SVM	least-squares SVM with gradient information
GPR	Gaussian process regression
GRU	gated recurrent units
GSSVM	grid search support vector machine
HDN	hierarchical diagnosis network
***K*NN**	*k*-nearest neighbors
LDA	linear discriminant analysis
LN	layer normalization
mRMR	minimum redundancy maximum relevance
ME	mixed eccentricity
MEISVM	multivariate integrated incremental support vector machine
MD	Mahalanobis distance
ML	machine learning
MLP	multilayered perceptron
MMD	the maximum mean discrepancy
MODBNE	multi-objective deep belief networks ensemble
MSA	multi-head self-attention
MSE	mean squared error
MSF	multiview sparse filtering

NLP	natural language processing
NSAE-LCN	normalized sparse autoencoder–based local connection network
IDDAN	inferable deep distilled attention network
IGBT	insulated gate bipolar transistors
IM	induction motor
IMF	intrinsic mode function
IMS	intelligent maintenance system
IRT	infrared thermography
PCA	principal component analysis
PCG-CNN	predictive consistency–guided convolutional neural network
PL	pooling layer
PPMCC	Pearson product–moment correlation coefficient
PSPP	Pythagorean spatial pyramid pooling
PWM	pulse width modulation
RBF	radial basis function
RBM	restricted Boltzmann machine
RDI	redefined dimensionless indicator
RMSE	root mean square error
RKHS	replica kernel Hilbert space
RUL	remaining useful life
SA	self-attention
SAE	sparse autoencoder
SE	static eccentricity
SF	sparse filtering
SGD	stochastic gradient descent
SOM	self-organizing map
SSA	singular spectrum analysis
STFT	short-time Fourier transform
SVD	singular-value decomposition
SVM-RFE	support vector machine with recursive feature elimination
SVR	support vector regression
TCA	transfer component analysis
TICNN	convolution neural networks with training interference
TL	transfer learning
***t*-SNE**	*t*-distributed stochastic neighbor embedding
UMP	unbalance magnetic pull
ViT	vision transformer
VSI	voltage source inverter

WDCNN	convolution neural networks with wide first-layer kernels
WPT	wavelet packet transform
WT	wavelet transform

Chapter 1
Introduction

1.1 Structure of electric powertrain in EVs

The motor drive system plays a significant role in the safety of electric vehicles as a bridge for power transmission. Meanwhile, to enhance the efficiency and stability of the drive system, more and more studies based on AI technology are devoted to the fault detection and diagnosis (FDD) of the motor drive system. This chapter reviews the application of AI techniques in motor FDD in recent years. AI-based FDD is divided into two main steps: feature extraction and fault classification. The application of different signal-processing methods in feature extraction is discussed. In particular, the application of traditional machine learning and deep learning algorithms for fault classification is presented in detail. In addition, the characteristics of all techniques reviewed are summarized. Finally, the latest developments, research gaps, and future challenges in fault monitoring and diagnosis of motor faults are discussed.

FDD is a condition detection technique that can be used to identify the local or overall operating condition of an electric motor, detecting early faults, and making predictions. The purpose of this technology is to detect faults and distinguish different types of faults to make decisions in advance to prevent the happening of hazards. In addition, modern industrial technology has been developed significantly, and electric motors have been widely used in different areas of industrial systems, especially playing a key role in the powertrain of electric vehicles. Figure 1.1 illustrates the basic architecture of an electric vehicle drive system. The conventional powertrain of an electric vehicle consists of electrical and mechanical systems. The electrical part includes the electric motor, the battery management unit, the power electronic devices, and the controller module, while the mechanical part includes the gearbox and the wheels.

Therefore, the motor is the most important component in the conversion of electrical energy to mechanical energy. The operational status and related parameters of the electric machine can be monitored and acquired in real-time through an online system. However, as the operating time increases, the possibility of motor failure under different operating conditions raises, threatening the reliability and safety of electric vehicles.

Electric motors suffer from several types of faults [1]. Generally, the two common forms of faults are the rotor and bearing faults in electric motors. The

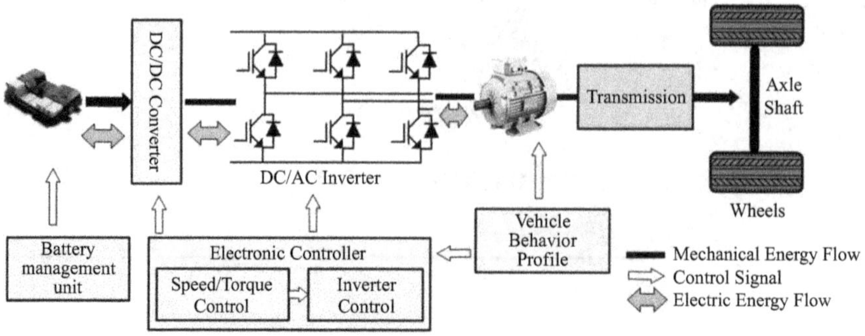

Figure 1.1 The basic structure of the motor drive system

failure of one component of a motor will lead to a chain reaction, which can fail or even paralysis of the entire equipment system [2]. Therefore, it has become imperative to enhance the safety and reliability of motor drive systems. The effective diagnosis of electrical or mechanical faults in electric motors is necessary [3–6]. For noninvasive motor FDD techniques, different signals can be analyzed, including temperature analysis, vibration analysis, infrared analysis, current and voltage analysis, electromagnetic field analysis, and ultrasonic analysis. Vibration analysis [7–9] and motor current signature analysis (MCSA) [10,11] are two of the most popular research in this field. Stator current analysis is known for providing noninvasive condition monitoring for EV motors [12–14]. Accurate detection of potential or existing motor faults is an essential measure to maintain safe machine operation. Among the theoretical methods and approaches that have been studied, the technical strategy of motor fault diagnosis in different external environments and operating conditions is a guarantee of improving the reliable operation of the equipment system, which will bring certain limitations [15,16].

In general, the motor FDD techniques are divided into model- and data-based methods. Model-based methods for motor fault detection are common in industrial applications. The simplest approach is based on input and output signal processing. If the output of the motor is outside the normal range of variation, the motor is considered faulty or about to fail. Mathematical models are used to describe the output amplitude, phase, frequency, and correlation with the source of the fault, and then these quantities are processed to determine the location and cause of the fault [17]. Moreover, classical methods based on state estimation or process parameter estimation are also commonly used for motor fault detection [18]. The advantage of these methods is that they can drill down in the basic dynamic characteristics of the motor system for real-time diagnosis, but an accurate mathematical model of the motor is required. Therefore, such methods are difficult to implement when the motor system model is uncertain or nonlinear. Figure 1.1 indicates the experimental rig for the application of motor drive system FDD techniques, which includes the motor, controller, load, sensors, and so on.

In recent years, the data-based approach has gained popularity due to its high practicality. It is a suitable method for incorporating artificial intelligence into

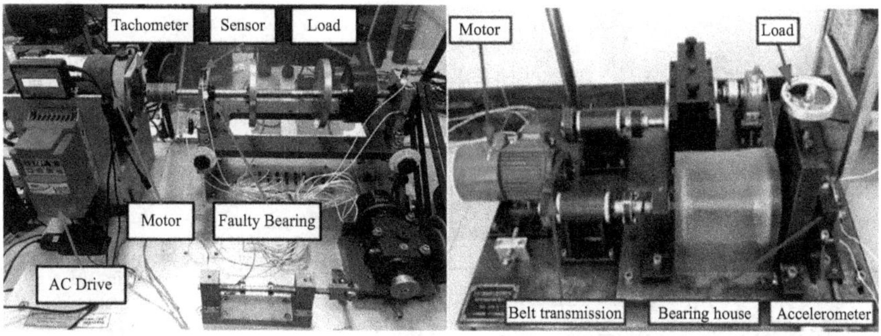

Figure 1.2 The experimental rig [42,106]

FDD. Figure 1.2 illustrates the general flowchart of FDD techniques. AI-based FDD is divided into two main steps: (1) feature extraction; (2) fault classification. In signal-based feature extraction engineering, the time-domain signal is usually converted to the frequency-domain signal by discrete Fourier transform. Simultaneously, time and frequency signal analyses become extremely important when the motor is under dynamic and transient conditions. Short-time Fourier transforms, Hilbert transforms, wavelet transform, and other signal-processing techniques have been widely proposed. Furthermore, time-domain FDD analysis is the best choice when using minimal computational resources to process extremely complex and intermittent signals. In addition, AI-based fault classification methods include machine learning, deep learning, fuzzy neural networks, genetic algorithms, and hybrid algorithms, which can address issues that cannot be solved by traditional fault diagnosis methods [19].

Artificial intelligence is the key point of data-driven technologies. Artificial intelligence–based studies have been extensively carried out in industry applications. Artificial neural networks (ANNs) use activation functions to predict interactions between artificial neurons. The related weights and biases are used to model the biological neural system. Nonlinear characteristics can be extracted and combined using ANNs, which can be employed to perform classification and regression tasks in the motor drive system. In [20], ANNs were applied to track and diagnose external faults in three-phase induction motors (IMs). The data of stator voltage, current, and motor speed were used to train the network. Furthermore, the identification and diagnosis of IM faults implemented in supervised and unsupervised neural networks have been achieved in [21], as well. The application of neural networks in the classification of partial discharges in motor insulation was studied in [22], where the fault of an IM is determined by the imbalance of the inductor current. Ref. [23] investigated the state estimation of IMs based on classical machine learning algorithm with nonlinear state estimators, which is based on stator current and rotor angular velocity measurements. In addition, DL can model and obtain accurate classification and predictions for complex fault types. Machine fault diagnosis has been effectively implemented using several common deep

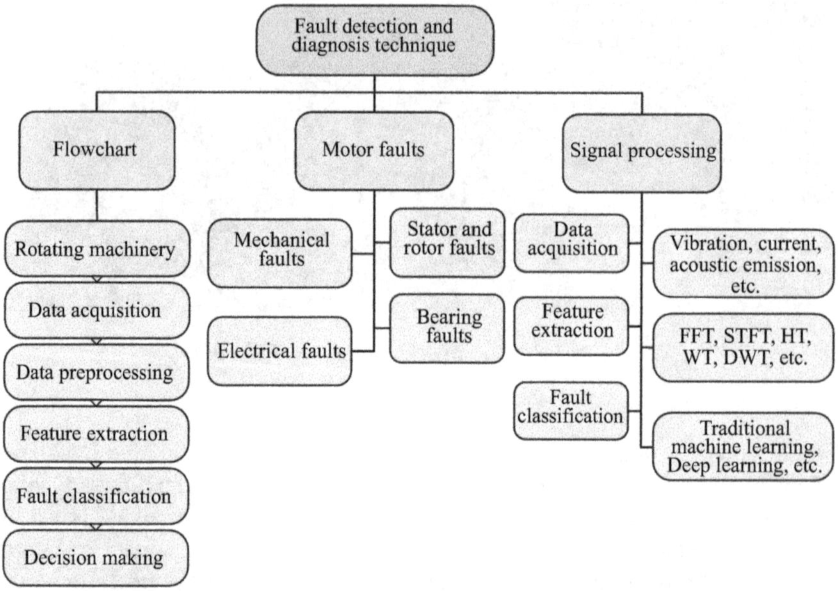

Figure 1.3 General flowchart of FDD techniques

architectures, including autoencoder (AE), recurrent neural network, generative adversarial network, convolutional neural network, and deep confidence network. Research on artificial intelligence–based techniques is significant for providing valuable research directions, but there is no comprehensive overview of the application of artificial intelligence–based techniques in motor FDD. It is important to investigate the contributions of different scholars in this field [332]. Figure 1.3 indicates the general flowchart of FDD techniques.

1.2 Review of faults in electric powertrain in EVs

In the advanced powertrain system, IMs and permanent magnet synchronous motors (PMSMs) are the two major types of electric vehicle motors. IMs are widely used in industry and are characterized by stator and rotor made of laminated silicon steel sheets and encapsulated with aluminum caps at both ends. Between the stator and the rotor, there are no mechanical elements in connection with each other, and the structure is simple with reliable operation. IMs are more efficient and appropriate for high-speed operation than direct current (DC) motors of the same power. The principle of operation is to generate an induction current in the rotor through the rotating magnetic field of the stator to produce electromagnetic torque, and no magnetic field is generated in the rotor; therefore, the rotor speed is less than the synchronous speed. Therefore, it is also called the asynchronous motor. Due to the complex mechanical structure of DC motors that restricts the overall vehicle performance, synchronous motors are gradually becoming popular drive motors. The rotor of PMSM produces a fixed direction magnetic field, and the stator rotating magnetic field drags the rotor magnetic field to rotate, so the rotor speed is equal to the synchronous speed. PMSM has the advantages of high efficiency, high starting torque, and high power factor.

Electric motors and drives are affected by many different types of faults, (1) stator faults can be classified as stator open phase, the short circuit that leads to stator imbalance or increased resistance connections. Loose electrical parts or poor connections can cause heat and eventually fire. (2) Rotor electrical faults, which include rotor open phase, rotor imbalance due to short circuits or increased resistance connections and broken bars or cracked end rings in IMs, and rotor magnetic failures such as demagnetization in permanent magnet synchronous machines. (3) Rotor mechanical faults such as bearings damage, bent shaft, eccentricity, and misalignment. (4) Faults of the power electronic components in the motor drive system. Figure 1.4 illustrates the two typical faults of the stator winding short-circuit and rotor crack fault. Listed next are the main categories of faults that can be detected with the MCSA. The frequency caused by each type of fault depends on the specific characteristics of the motor and the operating conditions.

Stator winding fault: Most winding faults result from the growth of uncorrected turn-to-turn defects. Long-term thermal aging and eventually insulation failure are the major causes of turn-to-turn defects. It may result in the opening, shorting, or grounding of one or more winding circuits, excessive heating, and machine failure.

Bearing faults: Bearing faults can be caused by several factors: (1) Due to the huge output load torque, there is a high vibration in the rotor, which leads to large fatigue stress; (2) incorrect bearing installation; (3) high currents in the bearing

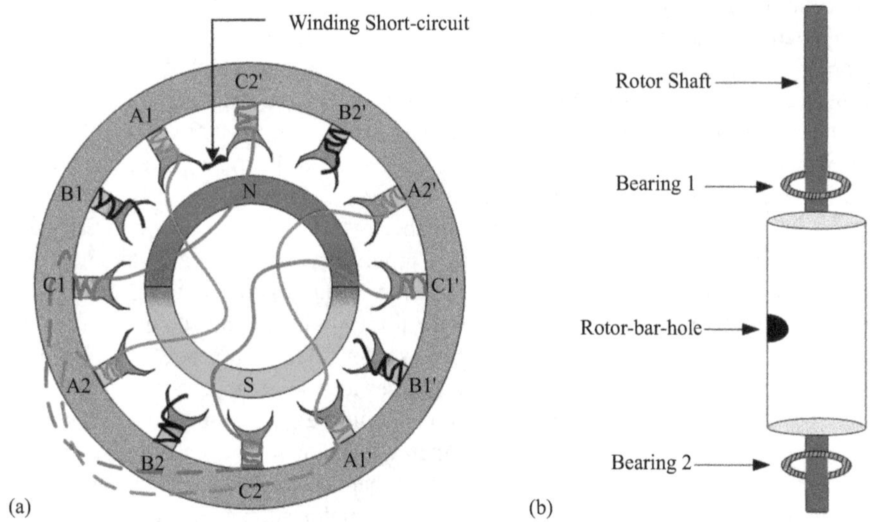

Figure 1.4 Typical faults: (a) stator winding short-circuit (WSC); (b) rotor crack fault (CRF)

directed by shaft voltage cause lubrication to deteriorate; (4) heat transferred to the shaft, resulting in friction and pollution. Bearing faults eccentricate the rotor, resulting in imbalanced magnetic forces and increased bearing stresses. Because the shaft dynamics are impacted by the distorted air gap between the stator and rotor, as well as variations in bearing stiffness, bearing failure is one of the reasons for excessive motor vibration.

Air-gap eccentricity occurs when the air-gap distance between the rotor and the stator is not uniform. The two types of abnormal air-gap eccentricity are static and dynamic abnormal air-gap eccentricity. In the case of static eccentricity, the location of the minimal radial air gap is constant, but in the case of dynamic eccentricity, the position of the minimal air gap follows the rotation of the rotor. When the rotor bars retreat or approach the stator magnetic fields, the current in the stator varies.

Broken rotor bar faults: Broken rotor bars are caused by faults in the rotor bars and end rings (BRB). Frequent starting under rated voltage, thermal imbalance, overload at the beginning (thermal stress), and unbalanced magnetic tension are all potential causes of BRB. These faults induce localized heating or arcing in the rotor, as well as vibration from the expansion and bending of the rotor.

1.3 Review of monitoring techniques for electric powertrain

MCSA is a condition-monitoring approach for diagnosing electric motor issues. The idea was first presented in the early 1970s for use with inaccessible motors in nuclear power plants and motors located in dangerous regions. It has rapidly gained recognition in the industry in recent years. Motors are operated under load and evaluated online without stopping production under regular operating circumstances. MCSA can be used as a preventative maintenance technique to identify typical motor faults before they become serious, which prevents costly catastrophic faults, and production interruptions and extend motor life. MCSA is an electrical characterization approach that may be used to assess electric motors, generators, power transformers, and other electrical equipment. The MCSA is used to detect faults by monitoring the stator current of the motor. Only one of the three-phase supply currents of the motor is typically monitored with a single stator current monitoring device. The MCSA employs the motor stator winding as a transducer to acquire signals (induced currents) from the rotor while also revealing information about the condition of the stator. A current sensor (clamp-on probe, current transformer) with a resistive shunt at its output senses the motor current and records it in the time domain. Ideally, the motor current should be a pure sine wave. However, the motor current contains numerous harmonics. The motor current signal is further modulated by the different electrical and mechanical faults present in the motor, resulting in extra sideband harmonics. Faults in motor components cause abnormalities in the magnetic field, affecting the mutual and self-inductance of the motor. In the motor supply current spectrum, they appear as sidebands around the line (power, grid) frequencies. Motor faults can be diagnosed, and their severity can be validated based on fault characteristics. Due to the frequency range of interest for MCSA is generally 0–5 kHz, Nyquist's theorem states that a sampling rate of at least 10,000 times per second is required.

MCSA is the best choice for motors under steady-state conditions and rated loads since the current changes are influenced by both the fault and the power supply. The impact of eccentricity and the bearing faults can be reduced using a rapid current-controlled inverter. Three effects are present in the faults: (1) the pulse repetition frequency, which is determined by the rotation frequency; (2) the vibration caused by the pulse; (3) a rise in the total noise level. The most accurate method is to use sensor signals to detect the existence of these faults. As the sensor is close to the fault location, the relationship between the fault and the sensed variable becomes clear. However, in addition to MCSA, many different fault diagnosis methods have been proposed.

One of the essential condition-monitoring approaches is oil and lubrication analysis. Online (particle counting, temperature and viscosity monitoring, etc.) and offline (oil filter analysis for flow and cleanliness characteristics) methods are both used to test and analyze lubricant samples. Lubrication analysis aims to retain oil quality and ensure that the components involved operate in the best possible environment. Lubrication analysis is generally done offline, with samples being examined and tested. Furthermore, when oil filters become overly dirty owing to component wear, they might be a reliable sign of faults. Acoustic emission analysis is carried out using sensors that capture the sound generated by the machine using a sound-level meter. The pressure levels and vibrations are converted into voltage signals using devices with antialiasing and high sampling rates. The types of data collected by acoustic emission analysis and vibration analysis are the same. The acoustic signal obtained is oscillatory. The acoustic signal characteristics vary depending on the faults of the rotating machinery. The noisy background might introduce additional components and impair the accuracy of fault identification of the monitored component, which is the major drawback of the AE condition monitoring method. The most widely used approach for condition monitoring is vibration analysis. Any change or malfunction in any of the mechanical components will cause the vibration profile to alter. Monitoring the vibration frequency can reveal whether a component is faulty or not. The disadvantages of vibration analysis are measurement errors due to improper sensor mounting, crystal overheating, and expensive proximity probes. Thermal field issues in motors are of great concern, and accurate temperature calculations are critical to the design and operation of motors. In addition, accurate thermal modeling of motors is essential for condition monitoring of the motors. Motor losses can result in high temperatures, which can cause severe thermal stress. There is thermal stress in the rotor end ring and bar, which is most likely to cause motor faults. Therefore, analyzing the effect of faults on the performance characteristics of the motor through the temperature rise of the motor can provide guidance to prevent accidents in operation and maintain safety. However, thermal analysis relies heavily on the accuracy of the sensors and measurements. The analysis of air gap or stray flux measurements can directly indicate the asymmetry of the radial or axial flux of the motor generated by fault-induced anomalies. Researchers have developed methods based on the analysis of external magnetic fields. Noninvasive examination and simplicity implementation are two of the main advantages. The drawbacks of these approaches stem from the complexity of simulating the magnetic field, which is strongly dependent on the electromagnetic behavior of the stator yoke and the motor housing, both of which have significant shielding effects. One of the main reasons that flux monitoring has not received as much attention as MCSA is because it fails to provide remote monitoring. However, it is a low-cost alternative method that can compensate for the reliability and variety of fault detection in the limitations of electrical, mechanical, and thermal monitoring. Initially based on the detection of modes associated with the current Park's vector representation, Park's vector approaches have been effectively used for condition monitoring of electromechanical systems. The extended Park's vector method relies on spectral analysis of the alternating current (AC) level of the current Park's vector mode, while by averaging the current Park's vector method, converter power switching faults are detected when the vector mode is not zero. The major risks of converter diagnostics are load dependency and susceptibility to transients. The phase currents are normalized by the modulus of Parks' vector, and the absolute value of the phase derivative of the absolute Park's vector is used as the detection variable in the power converter self-diagnosis based on Parks' vector technique. Multiple faults can be diagnosed with extra signal processing and variables.

1.4 Errors in the electric powertrain

1.4.1 The three-phase power inverter

The inverter, which represents one of the most used circuit configurations used for speed control in many three-phase machine applications, is a key component within any EV. Its main objective is to produce an AC output waveform from a DC supply. A fixed DC input voltage is given to the inverter and a controlled AC output voltage is obtained through the adjustment of the on and off periods of its interior components (switches). Being such a vital component, it controls the electric motor; and being analogous to the engine management system within gasoline cars, it determines the EV's driving behavior. In instances, when the brake pedals of the vehicle are engaged and the motor operates as a generator, the inverter acts as a rectifier and returns energy to the battery. High-power insulated gate bipolar transistors (IGBTs) or silicon carbide metal oxide semiconductor field effect transistors represent the key constituents of the power inverter that control the flow of energy to the motor to initiate motion. Both the top and bottom switches of the inverter leg are controlled via the microcontroller and gate drivers. As the motor operates, the current, position, and voltage signals are sensed and fed back to the controller to change the inverter's modulation. An efficient modulation algorithm, swift feedback, and precisely sensed signals are necessary for effective motoring to take place. Errors in the power inverter could occur due to inverter intermittent misfiring because of defects in control circuit elements or electromagnetic interference that results in gate-drive malfunctions [279].

To guarantee that the top and bottom switching devices never conduct simultaneously, a small time delay is added to the gate signal of the turning on device. This delay is termed dead-time. In other literature, it is referred to as blanking time or delay time. The dead-time is inserted when both the upper and lower switches within the leg of the inverter is switched off [8]. Its value depends on the type and power rating of devices used to construct the power inverter circuit [280–282]. For power semiconductor devices with notably high switching speeds, a proper dead-time must be selected to ensure the correct and safe operation of the inverter. Since the two active switching devices are in off-mode during the dead-time, either the top or bottom diode conducts depending on the direction of load current flow. Hence, an error could be generated in the output voltage depending on the direction of this load current. Even though the error in voltage is only for a short duration, cumulatively the impact of this could be significant especially when carrier frequency is high [283]. The insertion of dead-time assures safe operation, nevertheless, it causes major output voltage distortion, which further leads to an increase in current harmonics and deterioration in control performance [284–287]. Inverters have other nonideal factors such as voltage drops across the switches during on-states, switching delays, parasitic output capacitances of switches, and propagation delays in gate-drive circuits. These factors also serve as sources of output voltage distortion [288–291].

1.4.2 Analysis of dead-time effects

A typical configuration of a voltage source inverter (VSI) interfaced with a PMSM load is depicted in Figure 1.5, IGBTs are used as switching devices. The pole voltages (voltage at the midpoint of a leg measured with respect to the midpoint of the dc bus are referred to as V_{A0}, V_{B0}, and V_{C0}). If the three-phase VSI is modulated using sine-triangle PWM strategy, then the three sinusoidal modulating waves can be compared with high-frequency triangular carrier waves, and gating pulses can subsequently be generated for the six switches of the VSI. In the comparison, if the

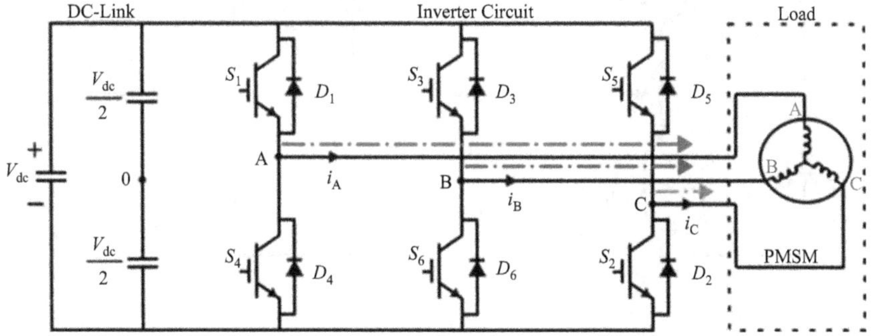

Figure 1.5 Three-phase PWM inverter with a PMSM load

modulating signal is greater than the carrier wave, the top device will be on; otherwise, the bottom switch will be active.

The effect of blanking time on pole voltage transitions: Depending on whether the load current is positive or negative, the pole voltage transition will take on different patterns and an error will always ensue due to insertion of the blanking time. The load current direction is considered positive when its direction is toward the load.

By referring to Figure 1.6(a), if the gating pulse of S_1 is high, it will conduct and current will flow from its collector to the emitter, then finally to the load. However, if the gating pulse to S_1 is low, it will not turn on. The only way for the current to continue flowing positively to the load is through the feedback diode D_4. On the other hand, the load current is considered negative when its direction is toward the inverter.

Consider the instance, when the gating pulse of S_4 is high as depicted in Figure 1.6(b), it will conduct and current will flow through it conventionally. However, if its gating pulse becomes low, even though the gating pulse of S_1 will be high at this instant, it will not be able to conduct due to the direction of the load current (i_A). Since current can only flow through the switch from collector to emitter, the only available path for the load current to flow is through feedback diode D_1. Hence in summary, it can be observed that whenever the load current is positive, either top switch S_1 or bottom diode D_4 will be turned on. Likewise, whenever the load current is negative, either top diode D_1 or bottom switch S_4 will be on.

The gating pulses (ideal and actual) as well as pole voltages (actual and ideal) are depicted in Figure 1.7 when the load current is positive. By considering Figure 1.7(e) for initial analysis, when S_1 is on and the load current is positive, the pole voltage in phase leg A (V_{A0}) equals half of the dc bus voltage $\left(+\frac{V_{dc}}{2}\right)$. Similarly, when D_4 is on and the current flow is toward the load, V_{A0} emerges as negative of half the dc bus voltage $\left(-\frac{V_{dc}}{2}\right)$.

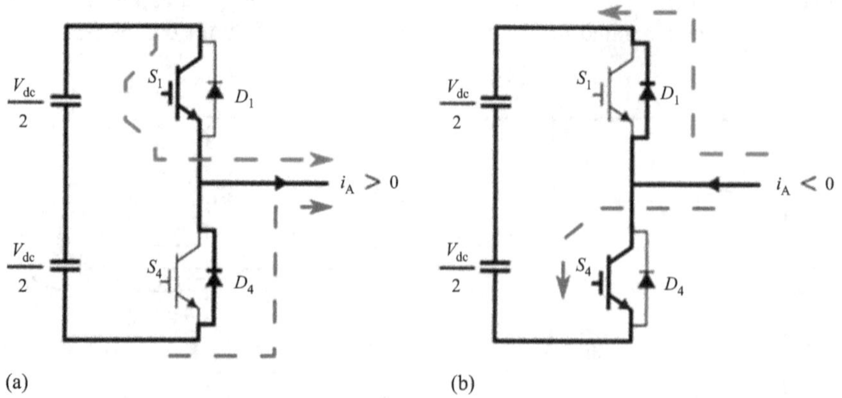

Figure 1.6 Impact of current flow on pole voltage transition for (a) positive direction of current flow and (b) negative direction of current flow

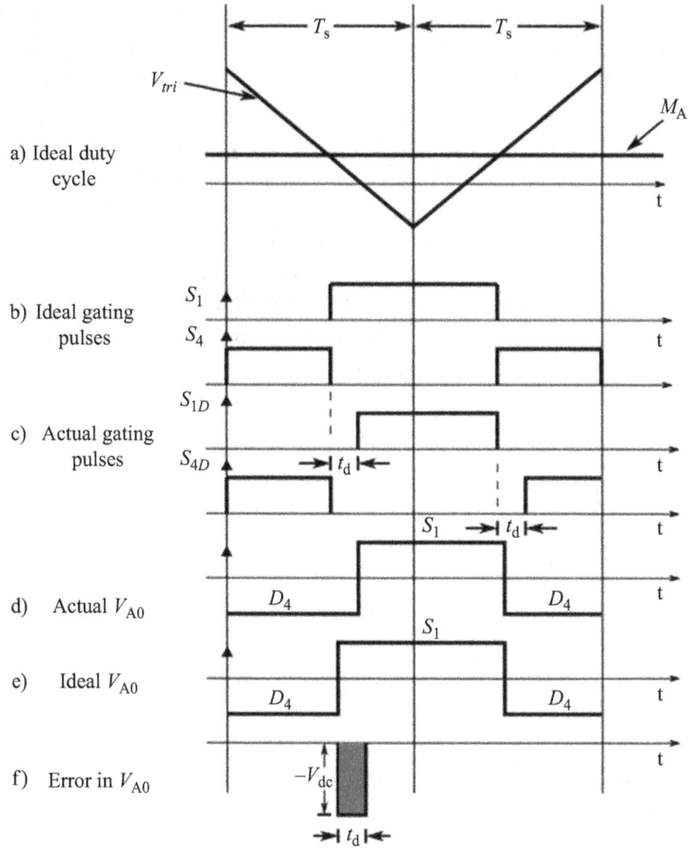

Figure 1.7 Pole voltage transitions for positive load current

From Figure 1.7(c), by inserting the blanking time, the rising edges of both actual gating signals are delayed by time, t_d. The actual pole voltage pattern is generated based on the behavior of actual gating signals. By referring to Figures 1.6(a) and 1.7 (c) when S_{1D} (actual gating pulse of S_1) is low, the complementary switch S_{4D} (actual gating pulse of S_4) happens to be high; however, since i_A is positive, it cannot flow through the bottom switch hence the feedback diode D_4 will conduct at a pole voltage of $\left(-\frac{V_{dc}}{2}\right)$ instantaneously. During the dead-time, the gating pulses of both S_{1D} and S_{4D} remain low; hence, it can be seen from Figure 1.6(d) that the diode continues to conduct during the delay to keep i_A in the positive direction, maintaining V_{A0} at $\left(-\frac{V_{dc}}{2}\right)$ in the process. Subsequently, the gating pulse of S_{1D} is made high and the current flows in the usual fashion to the load and conducts at $\left(+\frac{V_{dc}}{2}\right)$ pole voltage. When the gating signal of S_1 is made low, it immediately turns off. The rising gating pulse of the bottom switch is delayed by the dead-time. At this stage, pulses happen to be low for both switches, however, since i_A remains positive, the only device which will conduct to maintain the current in this direction is D_4 at a pole voltage of $\left(-\frac{V_{dc}}{2}\right)$.

A comparison of Figure 1.7(d) with (e) depicts that the introduction of the blanking time changes the appearance of the actual pole voltage pulse. A useful observation can be made in Figure 1.6(d), again in comparison to its ideal counterpart. It can be noticed that the transition pole voltage is delayed by the dead-time if the incoming device is a transistor, whereas it does not get delayed if the incoming device is a diode. The difference between the actual and the ideal pole voltages results in an error pole voltage of $-V_{dc}$ (Figure 1.7(f)). Hence in one carrier cycle, there is always one error voltage pulse in the pole voltage. The pattern will repeat itself in subsequent carrier cycles.

A similar analysis like was done previously is carried out in examining the pole voltage transitions when the load current flows toward the inverter. The ideal and actual pole voltages together with the gating pulses (ideal and nonideal versions) are depicted in Figure 1.8. By comparing Figure 1.8(d) and (e), just as was

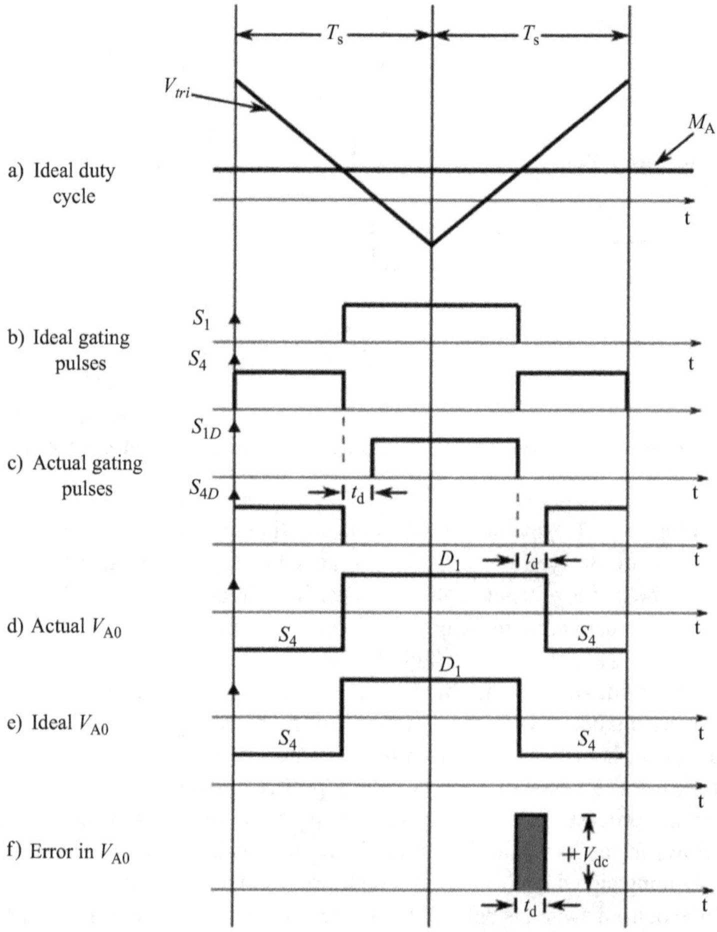

Figure 1.8 Pole voltage transitions for negative load current

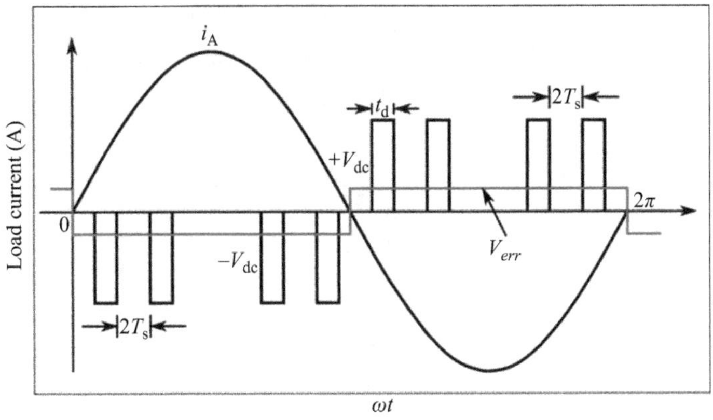

Figure 1.9 Error voltage pulses over a cycle of fundamental load current

done when i_A was positive, it can be observed from the generation of the actual pole voltage pattern that when the transition is from diode to transistor, the transitional pole voltage is delayed by t_d, however, when the transition is from transistor to diode, there is no delay. Therefore, in one carrier cycle when the load current is negative, an error voltage pulse of height V_{dc} and width equal to t_d is obtained.

Figure 1.9 depicts the generation of error voltage pulses over the positive and negative cycles of the fundamental load current. In the positive half cycle of the load current, negative pulses exist and vice versa. Since the pulses have the same height and width in every carrier cycle, they will have the same average value in all carrier cycles; hence, a square wave can be used for approximation. The average error pole voltage is given by

$$V_{err} = \frac{V_{dc} \times t_d}{2T_s} \tag{1.1}$$

where V_{err} represents the average error pole voltage, V_{dc} is the dc source voltage, t_d is the dead-time inserted, and T_s represents the switching period.

By applying the Fourier series to the square wave voltage, it can be decomposed into fundamental component (a small portion of it) and low-frequency portions (5th, 7th, 11th, and 13th) as well as the switching frequency components (f_s, $2f_s$). The troublesome components are the fundamental and low-frequency constituents.

The insertion of the blanking time ultimately introduces the average error pole voltage into the inverter output voltage. $V_{A0}(Av)$, the ideal average pole voltage, represents a scaled version of M_A. For the dead-time analysis for the zero-power factor instance shown in Figure 1.10(a), i_A lags $V_{A0}(Av)$ by 90°. When i_A is in its negative cycle, the average error pole voltage is positive and vice versa. Hence, the actual pole voltage is the sum of $V_{A0}(Av)$ and V_{err}. The actual pole voltage is no longer purely sinusoidal, as its distorted significantly.

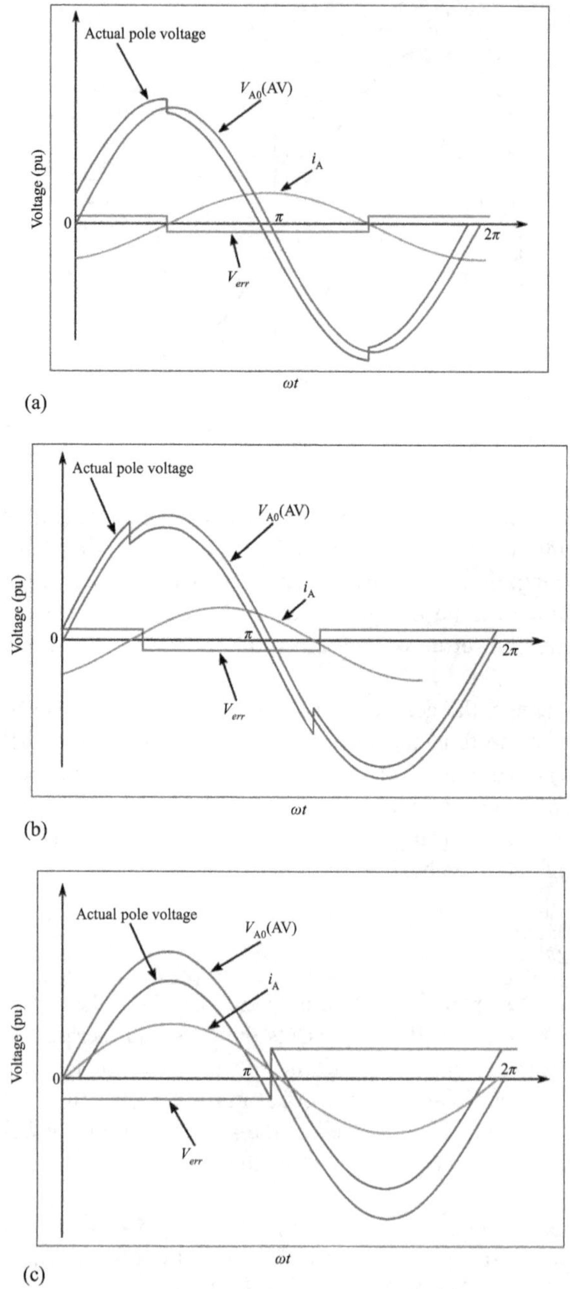

Figure 1.10 (a) Dead-time error for sine-triangle PWM technique for zero-power factor scenario, (b) for lagging power factor scenario, and (c) at unity power factor scenario.

In the lagging power factor case, i_A lags $V_{A0}(Av)$ by some angle below 90° as depicted in Figure 1.10(b). Again, the actual pole voltage is not purely sinusoidal due to distortion introduced by the dead-time.

Figure 1.10(c) portrays the unity power factor scenario, where i_A is in phase with $V_{A0}(Av)$. To obtain the actual pole voltage, $V_{A0}(Av)$ is added to V_{err} in all three scenarios and additionally the square wave is out of phase with the fundamental current in all three power factor scenarios. The average error pole voltage depends on the direction of the fundamental current.

The effect of dead-time on pole voltage transitions has been analyzed for positive and negative load currents as well as the generation of error voltage pulses over the positive and negative cycles of the fundamental load current. Furthermore, it has been pictorially shown how the insertion of blanking time leads to output voltage distortions.

Over the past few decades, substantial research has been carried out for VSI nonlinearity compensation in three phase motor drives. Numerous dead-time compensation strategies exist in literature today. The conventional dead-time compensation methods can be classified into two major categories. The first group, classified as pulse-based compensation strategies, detects voltage error introduced by the dead-time and then compensates for it in the next PWM cycle. The dependence on the current polarity is crucial in the pulse-based methods. In [287,292–296], current polarity detection is made possible with direct analog-to-digital (A/D) samples. However, these approaches are imprecise and exhibit inefficiency at detecting current polarity especially around the zero-crossing region. Consequently, if the polarity information suffers from imprecision, the distortion will worsen after compensation.

In [297], polarity detection precision is enhanced through feedback circuits to measure inverter output voltages and Kalman filters to perform the reconstruction of fundamental phase currents. A special detection circuit is used in [298] to detect the current polarity, with the VSI being regulated by a field programmable gate array. These approaches require extra circuits that increase the overall cost and hardware complexities. Additionally, majority of these methods are not completely accurate under entire operating conditions since the switching device parameters such as turn-on/off times, saturation voltage, and diode forward voltage change depending on operating conditions [289]. Hence, precise compensation cannot be guaranteed as the operation condition varies. All pulse-based techniques correct the output voltage magnitude accurately but possess a phase delay [299]. Consequently, phase lag elimination methods have been proposed in [300–302].

The second category comprises compensation strategies in which the voltage error caused by the dead-time is reduced by the alteration of a reference voltage. In these solutions, a compensating voltage is appended to the reference voltage [303,304]. These compensation techniques can be called average value compensation strategies or volt-second compensation schemes because they use the average value of a compensating voltage [305,306]. Even though the implementations of these methods are straightforward, they require precise current polarity and compensation time [290,292,307]. Majority of the schemes use current sensors to detect the polarity of the current [308]. However, due to the zero drift and the noise of current sensors, it is a challenge to judge the current polarity accurately especially when the current crosses zero. An approach, based on voltage sampling strategy to obtain the current polarity, is proposed in [309]; nevertheless it leads to disappointing results due to the fluctuation of neutral point potential when the methods are applied to three level inverters. In [289], an approach for instantaneous back calculation of current phase angle with a single current sensor is proposed, while the introduced time delay and phase error cannot be ignored. In [298,310], high precision current polarity judgment is realized by detecting which antiparallel diode conducts in a phase leg. However, the necessary hardware circuit is relatively complicated.

Although these strategies achieve good performance, they accrue cost due to the use of extra circuits. By comparing the two compensation schemes, the average value compensation scheme is easy to achieve albeit exhibiting a slow compensation action. The other category provides more precise and faster compensation but increases the overhead on the processor [280,291].

To bypass the problems of current polarity detection, an alternative approach based on disturbance observation is studied in [311–314] where the total voltage error is estimated through observers. These methods require accurate motor parameters and do not need current polarity detection and offline experimental measurements. It requires lots of experience and effort to acquire the adequate value of tunable parameters [315]. Moreover, only systems installed with speed sensors and at least two current sensors can adopt this method; hence, application is limited [309]. In [315–318], distance observers are used to calculate the voltage errors caused by VSI nonlinearity in the synchronous *d–q* reference frame. The voltage errors are subsequently compensated by adjusting the voltage references through a feedforward loop. Though the approach leads to successful results, the estimation errors can deteriorate the compensation performance due to motor parameter changes. In [319], the predictive controller is used to enlarge the control system's robustness against parameter changes, and the parameters within the observer can adaptively be adjusted; nevertheless, it requires information on the output voltage feedback.

In the PWM inverter, voltage distortion due to the dead-time effects produce sixth current harmonics in the *dq*-axes synchronous reference frame [320]. Based on this theory, numerous compensation strategies based on monitoring the current harmonics introduced by dead-time effects have been proposed. For instance in [284], a novel adaptive filter-based compensation scheme is used to suppress the dead-time effect by decreasing the sixth harmonic in the synchronous reference frame. The outstanding feature of the proposed strategy is the use of the positive and negative sequence components of sixth harmonics from *dq*-axes currents. By using the sequence information of the sixth harmonics, the compensated voltage can be estimated directly utilizing the voltage equation, hence bypassing the phase lag of proportional integral (PI) current regulators witnessed in [320]. The proposed method is superior again because in comparison with [321], which requires no additional information besides the rotor speed and *dq*-axes currents. Compared to the average-based and disturbance observer-based methods, the proposed strategy is robust to motor parameter changes since a feedback path with variable gains is designed based on the magnitudes of the sequence harmonics [284].

Since the dead-time insertion introduces low-order harmonics [322], it is possible to reduce these harmonics through the use of multiple resonant controllers (RSCs). Even though the RSCs can compensate for the low-order harmonics with efficacy, they have a tendency to trigger system resonance in instances where higher order harmonics (e.g., 11th- or 13th-order harmonics) are compensated in order to fully alleviate the dead-time [323]. A very promising solution for the harmonic suppression is to utilize a repetitive controller (RC) premised on the internal model principle [324]. This can suppress all harmonics (dead-time inspired harmonics included) without having to worry about resonance issues, so long as a good design algorithm is adhered to. As such, a scheme is proposed in [323] in which an RC is plugged into a proportional resonant-based fundamental controller to get rid of dead-time effects in grid-connected PWM converters. In comparison with the conventional dead-time compensation approaches, the RC compensates the dead-time harmonics as well as other low-order distortions with efficacy. Even though this strategy is effective in significantly minimizing the low-order harmonics, it does not consider other dead-time-related high order harmonics. To remedy this flaw, reference [325] proposes a dead-time compensation scheme based on a revised RC (RRC) which is installed in conjunction with current PI controllers. Unlike the other RCs, the proposed RRC uses finite impulse response moving average filter with a simple structure and is easier to implement utilizing commercial microcontrollers. It guarantees the tracking precision and its parameters can be kept fixed for myriad operating conditions [325]. Due to the nature of repetitive control, RRC has better robustness against motor parameter changes compared to disturbance observer-based methods. To add to this, the scheme does not only minimize the sixth harmonics as in many other compensation strategies but also deals with its multiples simultaneously.

In summary, dead-time insertion is necessary to guarantee safety in VSIs, although it distorts the output voltage and results in additional components of low-order harmonics as well as distortion in current waveforms. Several strategies have been presented to annul the impact of this delay. The majority of the compensation schemes hinge on the polarity of the current and compensation voltage. Detection of current polarity when the phase current is close to zero continues to be elusive and acts as a major challenge as far as dead-time effects compensation techniques are concerned. To avoid the unwanted current distortions, caused by imprecise detection of current polarity, some authors have suggested the possibility of ignoring the compensation when the phase current is close to zero, and accurate polarity detection becomes indeterminate. This stance, nevertheless, could dent the integrity of the scheme. Since the voltage drops across switches influence the modulation index, especially at low current scenarios [326], some authors believe that to accurately compensate for the dead-time effects, the nonideal characteristics of switching devices must be considered. However, indices, such as turn on/off times, voltage transition slope, and on-state voltage drops, are not exactly measurable and are dependent on the operating point [282]. Consequently, in [326], the knowledge of the switching device properties is assumed. This assumption subjects the accuracy of the output voltage compensation process to lots of compromise. Evidently, absolute resolution of dead-time effects in VSIs remains an existing dilemma and to get to the root of the issue, the zero-crossing clamping effects need to be exhaustively studied and understood more. Again, the unpredictability of semiconductor switches under myriad operating conditions needs to be thoroughly comprehended for dead-time compensation schemes to be comprehensive in their pursuit.

1.4.3 Electric machine (motor)

Another crucial component of the EV powertrain is the electric machine (motor). Its primary objective is to convert electrical energy into kinetic energy to power the wheels. It plays a pivotal role in achieving the best possible drive range, just as the battery does. A good drive range has a direct link to excellent motor/machine efficiency (in essence how good the motor is at converting electrical energy into mechanical energy). Some of the very popular electric machines used in EVs are PMSMs, three-phase AC IMs and switched reluctance motors (SRMs). PMSM is the best choice for high performance motor drive applications due to its higher ranking efficiency, compact nature, and being light weight. Though PMSMs have attracted lots of attention over the last couple of decades due to superior properties, they are not invulnerable to various errors of electrical (stator windings), magnetic (permanent magnet demagnetization), or mechanical (bearings, imbalance, and eccentricity) nature caused by various stresses witnessed during lengthy operation in severe conditions. Due to such harsh exposure, electrical machines are susceptible to exhibiting abnormal behavior at some point in their operational lives. Errors, depending on their level of accumulation or severity disturb the motor's normal operation, producing several unwanted symptoms like unbalanced line currents, torque and speed pulsations, decreased efficiency, excessive heating, and increased losses. As a result of the increase of PMSMs in safety-critical applications such as EV motor drives, detection, identification, and isolation of errors or deviations at their incipient stages is crucial. Undetected deviations from acceptable performance norms can lead to high substitution costs or possibly entire system failure. Hence the establishment of monitoring systems plays a huge role in detecting and diagnosing flaws within drive systems. Monitoring errors or abnormal behaviors of the PMSM drivetrain involves surveilling of motor-related operating parameters such as current, voltage, flux, speed, temperature, vibration, noise, partial discharge, and the use of the correct strategy to evaluate the current operating state of the motor/machine. If the motor deviates from acceptable operational limits, it is necessary to further figure out the type of anomaly, location, severity, and development pattern [327]. These deviations from normal operation can seriously affect the reliability and safety of the motor operation and can lead to high torque ripple generation, low efficiency and other grave safety-related problems if intervention (timely diagnosis and remedial measures) is not taken.

In a healthy machine it is expected that the center of the stator (c_S), the center of the rotor (c_r), and the center of rotation (c_ω) all align or coincide. The misalignment of the geometric centers of the stator and the rotor due to manufacturing imperfections lead to eccentricity errors which rank as some of the major root cause of PMSM problems. This phenomenon gives rise to the asymmetry of air-gap length, resulting in the unbalance magnetic pull (UMP) between the stator and rotor. The UMP may further cause vibrations, noise, and mismatch in phase currents [328]. Generally, up to 10% eccentricity is permissible. When eccentricity becomes significant, it can cause rotor-to-stator rub and ultimately damage the rotor and stator winding [329,330]. Electric machine errors emanating from eccentricity can be categorized as static eccentricity (SE), dynamic eccentricity (DE), and mixed eccentricity (ME).

In SE, both centers of the stator and rotor do not coincide. In this case, the position of the minimal radial air-gap length is fixed in space. SE may be caused by the inaccurate position of the stator or rotor during the commissioning stage or simply by the stator-core ovality. Inherently, a small degree of SE tends to exist even in newly manufactured machines due to unavoidable manufacturing errors. DE occurs when the center of the rotor rotates around the center of the stator. In other words, in DE, it is the center of the rotor that is shifted from the center of rotation. This deviation may be caused by several factors such as bearing wear or misalignment, bent rotor shaft, mechanical resonance at critical speed [331]. Both rotor and stator centers are shifted from the center of rotation in the ME case. Stator current-based approaches are some of the most widely used strategies employed for detecting the eccentricity errors shown in Figure 1.11.

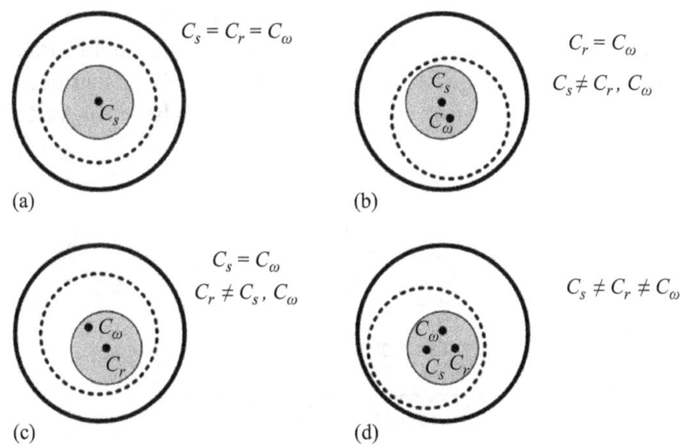

Figure 1.11 Typical PMSM exhibiting different types of eccentricity: (a) healthy case, (b) SE, (c) DE, and (d) ME

1.5 Analysis of EV powertrain errors

The electric powertrain must be extremely reliable since it is at the heart of all EV systems. For effective closed-loop control of the motor drive, the various drivetrain parts efficiently interact to enable smooth electric thrust. However, as mentioned earlier, errors of varying degrees emerge during the EV's lifetime to disorganize the operational cohesion within the powertrain components.

The microcontroller, power inverter, and PMSM depicted in the closed-loop control diagram rely on accurate electrical and mechanical signals from sensors as feedback. The errors that materialize consistently breach the communication pattern between these devices and aim to reduce the intended lifespan of EVs.

1.5.1 Contribution of errors from the PMSM

For the rotating machines, the radial electromagnetic forces are balanced when the rotor is concentric with the stator. Nevertheless, in practical systems, rotor eccentricity appears for several reasons such as manufacturing defects, assembly error, and bearing wear. This leads to unbalanced magnetic pull (UMP) that compels a shift of the rotor axis from the center of the stator, distorting the air-gap distribution. Consequently, UMP manifests a negative stiffness and nonlinearity effect causing problems within the dynamics of the powertrain. The unbalanced forces are highly dependent on the geometry of motors, especially on the structure of the stator teeth. The electromagnetic force acting on the surface of the stator teeth is a primary source of vibration and noise. Generally, vibration, which leads to acoustical noise, is a key index of a motor's performance.

1.5.2 Contribution of errors from the power inverter

In the PWM-inspired VSI, the output voltages consist of discrete pulses, the average values of which are equal to the reference over an entire switching period based on the voltage–seconds principle. Nonetheless, due to the nonlinearity of VSIs, there exist voltage distortions or error voltages between the actual output voltage and the reference voltage during a full PWM cycle. The error voltage occurs

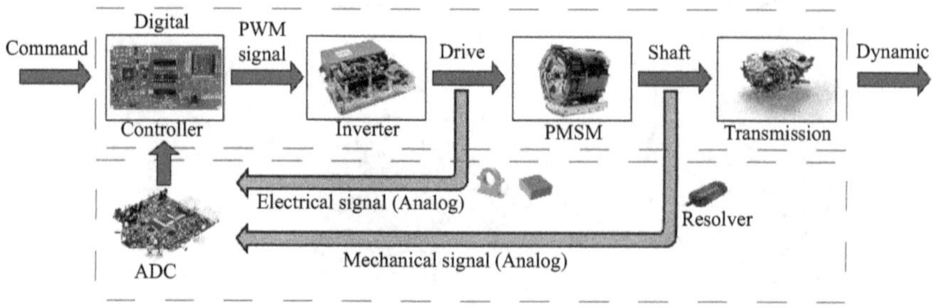

Figure 1.12 Scrutiny of the effect of errors within the closed-loop control of the drivetrain

because inverters inherently possess nonideal factors such as voltage drops across the switches during on-states as well as propagation delays in gate-drive circuits. Additionally, the dead-time is added to the gate-drive signals as a delay to avoid the Shoot-through phenomena.

Consequently, the effective duration time of an active power switch is affected by the dead-time as well as nonideal properties of power devices that result in voltage distortion. The VSI nonlinearities lead to torque pulsations, current distortions, a range of low-order harmonics and the deterioration of drive performance (especially in low-speed instances).

1.5.3 Contribution of errors from sensors

The outputs of the sensors for the closed-loop control system are connected to the controller since they cannot operate independently and need to be connected to an electronic device. They require to be in a healthy state to function properly. The mechanical and electrical feedback sensors, which represent some of the chief sensors used in the EV drivetrain, are subjected to myriad forms of imperfection because of noise and other cyclical errors.

1. **Mechanical sensors:** The accuracy of position measurement achieved by resolvers mostly depends on the quality of the analog signals and on the resolution of the RDC. Practically, no resolver generates ideal signals; thus, it is extremely difficult to meet precision specifications of an RDC. Resolvers generate periodic position errors due to nonideal output signal characteristics such as amplitude imbalance, quadrature error, inductive harmonics, excitation signal distortion, reference phase shift, and disturbance signals. These errors appear in the speed output of the RDC as oscillations of frequencies that are integer multiples of the resolver's mechanical frequency of rotation and are of magnitude proportional to the resolver speed.

   Ultimately, these position sensor errors lead to the generation of torque ripples, speed fluctuation, and unbalanced three-phase currents within the control loop.
2. **Electrical sensors:** These are used in the drivetrain to provide critical information pertaining to motion control, operating efficiency, and safety, which makes their performance and reliability very important.

   To ensure that the motor operates at its peak efficiency, its operational characteristics that are fed back into the control algorithm are critical. Phase current is one of such crucial diagnostic feedback elements used by the system controller to enable optimum performance. The quality of the current measurement (detected by current sensors) could directly determine phenomena such as torque ripple and torque settling time. To regulate the torque of the motor, the microcontroller requires instantaneous information on the phase currents of the motor in every control cycle. Two major errors which plaque the phase current feedback system with massive torque ripples are the offset and scaling error. Their impact is detrimental to the health of the motor drive.

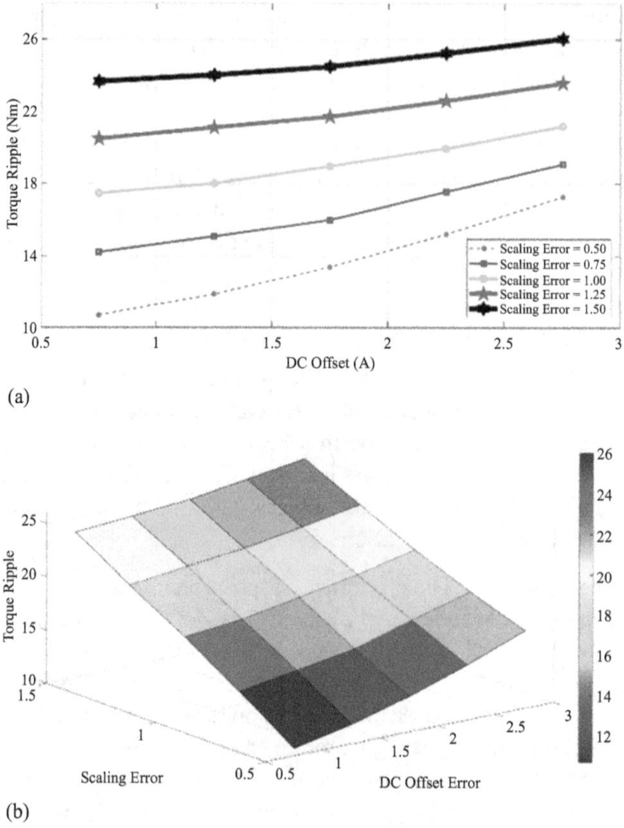

Figure 1.13 Torque ripple representation: (a) two-dimensional depiction versus DC offset, with a variation of the scaling error, (b) three-dimensional rendition with the DC offset and scaling error

Table 1.1 Parameters of simulation platform

Parameter	Value
Number of pole pairs	3
Resistance of stator	0.7917 Ω
d-Axis inductance	0.0595 H
q-Axis inductance	0.0595 H
Rotor flux	0.1796 Wb
DC voltage	100 V

An analysis was conducted in Simulink to investigate how the introduction of offset and scaling errors results in torque ripple generation.

The plots were obtained from an SPMSM-inspired model predictive current controller. Table 1.1 lists the parameters of the machine used for the simulation.

In using multiple current sensors for phase current measurement at the inverter output leads, the inherently different composition of each sensor leads to measurement mismatches, contributing to torque ripple generation. On the other hand, in the use of the single dc-link sensor, phase current reconstruction strategies employed result in considerable distortion in the current waveforms (due to modifications of the applied PWM voltage vectors leading to errors and performance deterioration within the powertrain).

1.5.4 Contribution of errors from ADC/microcontroller

Just like many other semiconductor components, high-speed ADCs do not always perform perfectly as anticipated. Their performance within the microcontroller has significant influence on the entire control loop. This performance is gauged by comparing the quality of the digital signal produced by the ADC to that of the original analog signal. A desirable feature in the ADC is to have the constituents introduced by the ADC during quantization as close as possible to their theoretical minimum limits. These quantization artifacts cause impairments to the precision of the digitized signal and may reflect as noise and distortion.

ADC errors can be categorized as those related to itself (offset error, gain error, full scale error, differential nonlinearity, and integral nonlinearity) as well as those related to the external environment (power supply noise, temperature impact, and analog signal source resistance). Both the offset and gain errors are constant errors and can be eliminated by appropriate calibration. The offset and gain error drift of the ADC contribute to system errors such as torque ripple, which lead to vibrations and excessive wear in machines. The DNL and INL on the other hand are difficult to calibrate and manifest themselves as noise within the closed-loop. There is also the issue of aliasing, which occurs when an input signal frequency exceeds the Nyquist frequency and it is replicated at other positions in the spectrum on either side of the Nyquist frequency.

In summary, this chapter has discussed some of the sources of error within electric powertrain and their overall impact on the motor drive. These errors emerge to shorten the EV's lifespan and need to be quickly detected and countered at all cost.

Chapter 2

Feature extraction engineering (FEE) for EV electric powertrain fault diagnosis

The basic task of feature extraction is to find out the most effective features for fault recognition from plenty of features, achieving the compression of the feature space dimension, that is, to obtain a set of "fewer but more precise" classification performances with a low probability of classification error. Motor current monitoring is an effective fault detecting technique since the stator current waveform data obtained from a defective motor differs from that collected from a standard motor. Most research on this technique uses different algorithms to decompose and interpret stator current waveforms, including Fourier analysis, linear discriminant analysis (LDA), wavelets, neural networks, and other predictive analysis approaches [24–27]. Time and frequency information is used in time–frequency domain strategies to capture transient characteristics [28]. Envelope analysis is a common technique for detecting and diagnosing bearing faults, which is traditionally determined based on analytically determined peaks. The fast Fourier transform (FFT) method is effectively implemented in the spectral analysis of the envelope signal [29]. However, the Fourier spectrum of vibration signals generally excludes the descriptive time-varying patterns of the acquired signals. Therefore, FFT fails to meet fault diagnosis requirements in real-time application [30]. In addition, most methods analyze vibration signals in the time or frequency domain. Vibration monitoring is considered a reliable approach to assess the overall health of a rotor system. The frequency-domain analysis is attractive due to the more detailed information provided about the machine status [31]. However, fault detection is performed by comparing the indirect measurements of external forces based on the dynamic behavior of the machine. The difficulty in fault detection lies in classifying many frequency lines present in the vibration spectrum to extract useful information related to the health status of the motor. Several studies have used dynamic signal analyzers to measure the variance of the spectrum over time to solve this issue [32]. Reference [33] developed an intelligent FFT analyzer that selects multiple frequencies in the spectrum as features and generates a reference model under healthy conditions, which is then compared to the monitoring characteristics for fault detection.

Furthermore, the Hilbert transform (HT) is presented as a method for extracting and estimating the envelope of the vibration signal to get the local energy at each instantaneous frequency. As a result, this technique may be used to characterize the energy–frequency distribution of the vibration signal. It is beneficial to extract the characteristics of nonlinear signals [34,35]. Due to the adaptive and unexpected nature of vibration signals, conventional methods must be based on reliable motor models and cannot be used effectively for vibration signal diagnosis.

The short-time Fourier transform (STFT) is a popular signal feature extraction approach for converting quasi-steady-state vibration data into a continuous spectrum for neural network model training [36]. Faults can be detected based on the change in the expected value of the vibration spectrum modeling error. However, many potentially unstable frequencies must be tracked since there is no precise method exists for predicting which form of failure will occur. Moreover, vibration spectra often contain puzzling combinations of unusual frequencies that provide little descriptive information about the operating conditions of the motor. Therefore, modern procedures for tracking motor conditions are not reliable or efficient enough. The recent performance of neural networks in dynamically modeled complex systems holds the promise of reducing these issues and achieving better fault detection performance [37]. Neural networks can describe any nonlinear model without knowing the exact form, returning fast results during the recall process. Combined with STFT, an analysis method using vibration spectrum neural network modeling was developed in [38] to extract fault spectrum features for detecting machine faults. For nonstationary processing signals, the wavelet transform (WT) is a useful approach [39], which has excellent time–frequency characteristics in the local area. The internal generation between the studied signal and the intended wavelet basis will provide detailed information in both time and frequency domains. Due to its multi-resolution inspection capabilities, this approach has shown great ability in fault diagnosis of mechanical machinery [40].

The discrete WT (DWT) was presented to extract fault features [41]. The advantage was to reduce the computation time. DWT has been widely applied in motor mechanical fault detection [42,43] since it can only decompose low-frequency sub-bands. Wavelet packet transform (WPT) was proposed to decompose the high and low-frequency band in parallel to improve its frequency resolution. Although these methods have achieved great performance, they were still limited by the segmentation scheme. WPT cannot split the frequency of the signal, which may severely damage the transient vibration characteristics [44].

To overcome the dichotomous subdivision scheme and achieve adaptive representation, the empirical modal decomposition (EMD) approach was suggested [45]. The multi-modulated vibration signal was decomposed into multiple intrinsic mode functions (IMFs) that were regarded as dominant mode components. It can extract both stationary and nonstationary components of a signal with efficiency. Therefore, it has attracted a lot of attention in signal processing and practical industrial applications. However, the lack of a mathematical theory for EMD methods has led researchers to propose improved methods regarding combinations of other methods in recent years [46]. Reference [47] proposed an empirical WT (EWT), which combined the advantages of WT and EMD methods. The EWT method can perform the identification of weak faults and compound faults. Meanwhile, the EWT method can efficiently analyze the signal and extract internal features. In addition, EMD was an adaptive signal-processing method that can be perfectly applied to nonlinear and non-smooth processes. The main drawback was the pattern mixing problem. In addition, a new method called variational mode decomposition (VMD) was proposed in [48], which assumed that each extracted pattern has a finite bandwidth. Compression is performed around the central frequency of the match. However, VMD cannot support the model in practical applications, and its modulation capability depends heavily on the intrinsic parameter settings [49].

Intrinsic time scale decomposition (ITD) is a new adaptive time–frequency analysis approach that allows a non-smooth signal to be decomposed into the sum of several intrinsic rotational components. Although it has better performance over EMD, the approach cannot account for the physical meaning of the intrinsic rotational components. Reference [50] developed a local characteristic scale decomposition (LCD) algorithm based on the ITD method for the physical significance and its intrinsic rotational components. It can decompose any complex signal adaptively into instantaneous physical frequencies [51,52].

Compared with EMD, the LCD method has better time–frequency localization characteristics and can extract the localization information of the original signal more effectively to obtain the fundamental characteristics of the signal. Since raw feature vectors will reduce the efficiency of the classification system, and incorrect generalization will eliminate useful details, linear methods are usually applied to reduce the number of features.

Both principal component analysis (PCA) and LDA can be used for 2D downscaling, achieving clear visualization and detection of different or unknown fault modes [53]. Many researchers have investigated PCA approaches, considering their drawbacks in working with massive data sets when it seeks the global structure of data [54]. The function vector is made up of D-computed features and has been expressed in D-dimensional space. Most of the data in the D-space has a nonlinear structure. In recent years, several learning methods [55,56] have been carried out to retain the information in a lower d-dimensional space, where $D > d$, to solve this problem. Other methods such as curvilinear component analysis (CCA) "find" the right form of the sub-tow cable and vice versa automatically. CCA has been applied in the field of fault diagnosis of motor drive systems. A CCA-based neural network FDD was presented in [57], which started with the selection of the most important features from an initial set formed by computing statistical time features from vibration signals.

The novelty of the method lies in its ability to perform complete fault analysis and diagnosis of various faults in bearings, both local and generalized faults. It is applied to the fault classification structure by performing the feature approximation phase of CCA. The classification task can be accomplished by applying a neural network.

Symmetric component (SC) is a promising method in time-domain FDD analysis and has received much attention, which can provide various information about the motor voltage or current balance, magnitude, and sequence. However, the generality of SC is limited to different fault classes, which may require detailed machine models and may be limited by computer resources. The SC method presented in [58] was based on stator currents and accomplishes signal feature extraction in multiple data-processing steps. It is a low-cost computational method without complex computer models. Furthermore, particle filtering is another efficient tool for sequential signal processing that uses point masses and related discrete probability masses to estimate the state probability density function. It is based on the principle of sequential value sampling and Bayesian theory. Particle filtering has recently been extended to the prediction of machines [59]. Since mechanical failures are nonlinear dynamic problems, particle filtration is particularly effective in solving these problems. In addition, some mathematical models have been developed to explain the fault propagation process in most applications. However, the derivation of these models is complex and requires expertise in the degradation process. The typical fault propagation model is the first-order hidden Markov model (HMM), which is characterized by the fact that the current state of the system, which is determined by the previous state. Moreover, due to the increase of feature nodes, the trained system may contain some redundant nodes, which will lead to poor accuracy. Singular value decomposition (SVD) has been proposed to simplify the system. In the SVD process, the signal is converted into a matrix where the singular values represent the nature of the faulty signal [60,61]. In addition, to fuse the information from all the utilized classifiers, reference [62] proposed a fault diagnosis method based on a motor current signal developing deep learning and information fusion (IF). The novel decision-level IF technique applied the raw signals of multiphase motor currents as direct inputs and extracted features from the motor current signals of each phase to improve classification accuracy. Table 2.1 summarizes all mentioned signal process methods for feature extraction.

2.1 Fast Fourier transform

The FFT is a fast and efficient algorithm for computing the discrete Fourier transform (DFT). Finite-length sequences occupy an important place in digital technology, mainly because their spectrum can be discretized. The DFT of a finite-length sequence can itself fully express the spectrum of that sequence, so the DFT can also directly analyze the spectrum of a signal. Spectrum analysis is used in communication technology, image transmission, speech compression, biomedicine, and other fields. Although spectrum analysis and DFT calculations are very important, DFT has not been used for a long time due to its high computational effort. Spectrum analysis was mostly solved by analog signal filtering and the situation changed radically in 1965 when two scientists from IBM in the USA, Cooley, and Tukey first proposed a fast algorithm for DFT computation. People began to recognize some of the intrinsic laws of DFT computation, which led to its rapid development and improvement. The advent of the FFT algorithm, a high-speed and efficient method of computation, greatly simplified DFT operations, reducing the computation time by 1–2 orders of magnitude. The advent of the FFT has led to the widespread use of DFT in practice.

Table 2.1 Highlight of feature extraction method

Reference	Theme	Principle	Highlight
[28–30,33]	FFT	• Spectrum analysis of envelope signals • Frequency distribution	• Time–frequency domain conversion • Poor performance for nonstationary signal • No time–frequency analysis
[32,34,35]	HHT	Multi-resolution of frequency scale	• To avoid complex mathematical operations • To analyze signals whose frequency varies with time • High calculation burden
[36,38]	STFT	Long- and short-time window movement	• Time–frequency characteristics are acquired • High-frequency resolution and high-time resolution cannot be acquired simultaneously
[39]	WT	Signal decomposition and reconstruction	Time–frequency analysis
[41,42]	DWT	Discrete input and output	To reduce calculation time
[43]	WPT	Fitting the mutation signal	Effective splitting of fault-induced resonances and high-frequency band features
[45,46]	EMD	Irregular frequency waveforming	• Adaptive data processing • Suitable for nonlinear, nonstationary time series
[48,49]	VMD	Subsignal decomposition at different frequencies	• To avoid pattern mixing • Modulation capability dependent on parameter settings
[47]	EWT	Adaptive wavelet subdivision scheme	More consistent decomposition
[44,50,51]	LCD	Adaptive decomposition of a complex signal	• Better time–frequency localization characteristics • More efficient extraction of localization information from the original signal • Faster calculation speed
[53,54]	PCA	Multidimensional feature mapping	• Signal dimensionality reduction • To reduce model training time • Better accuracy
[57]	CCA	• Information Retention • Self-organizing mapping	• Linear feature extraction • Distance preservation
[58]	SC	• Symmetric component • Acquisition	Rapid and accurate computation performance
[60,61]	SVD	Matrix decomposition	To use low-rank approximations to simplify systems

2.1.1 The calculation amount of DFT

The computational effort required to perform a DFT operation on a finite-length sequence (n):

$$X(k) = DFT[x(n)] = \sum_{n=0}^{N-1} x(n) w_N^{nk}, \qquad k = 0, 1, \ldots, N-1 \tag{2.1}$$

Generally, $x(n)$ and w_N^{nk}, $X(k)$ are complex numbers, so the DFT is computed as

$$\begin{bmatrix} X(0) \\ X(1) \\ \vdots \\ X(N-1) \end{bmatrix} = \begin{bmatrix} w_N^0 & w_N^0 & w_N^0 & \cdots & w_N^0 \\ w_N^0 & w_N^{1\times 1} & w_N^{2\times 1} & \cdots & w_N^{(N-1)\times 1} \\ \vdots & \vdots & \vdots & \cdots & \vdots \\ w_N^0 & w_N^{1\times(N-1)} & w_N^{2\times(N-1)} & \cdots & w_N^{(N-1)(N-1)} \end{bmatrix} \cdot \begin{bmatrix} x(0) \\ x(1) \\ \vdots \\ x(N-1) \end{bmatrix} \tag{2.2}$$

- Calculating the value of an $X(k)$ requires N multiplications of complex numbers and $N-1$ additions of complex numbers.
- Calculating the value of an $X(k)$ for N points require N multiplications of complex numbers and $N(N-1)$ additions of complex numbers.

In operations, multiplication is more complex than addition, especially the operation of multiplying complex numbers, where each multiplication of a complex number includes the multiplication of four real numbers and the addition of two real numbers, for example,

$$X(k) = \sum_{n=0}^{N-1} \left\{ \begin{array}{l} \left(R_e[x(n)]R_e[w_N^{nk}] - I_m[x(n)]I_m[w_N^{nk}]\right) \\ +j\left(R_e[x(n)]I_m[w_N^{nk}] + I_m[x(n)]R_e[w_N^{nk}]\right) \end{array} \right\} \tag{2.3}$$

Each addition of a complex number involves adding two real numbers, so each calculation of $X(K)$ requires $4N$ real number multiplications and $2N + 2(N-1) = 2(2N-1)$ real number addition operations. Therefore, the entire DFT operation requires $4N^2$ real number multiplications and $2N(2N-1)$ real number addition operations. Based on the previous analysis, in DFT calculations, the computational workload is proportional to N^2, regardless of whether it involves multiplication or addition. Hence, when N is large, the computational workload becomes significantly substantial.

For example, in a DFT with $N = 10$ points, 100 complex multiplications are required, while in the case of $N = 1,024$ points, more than one million (1,048,576) complex multiplications are required. Real-time processing requires high computational speed. The structure of the inverse transform (IDFT) and DFT operations are the same, except that they are multiplied by a constant $1/N$. Therefore, the computational effort is the same for both.

2.1.2 Operation characteristics of DFT

By examining the operational properties of the DFT, it was found that the amount of computation could be reduced by using the following two properties:

The coefficient $w_N^{nk} = e^{-j\frac{2\pi}{N}nk}$ is a periodic function, using its cycle periodicity and symmetry can improve operations and increase computational efficiency.

- Periodically:

$$w_N^{nk} = w_N^{n(N+k)} = w_N^{k(N+n)} \tag{2.4}$$

- Symmetry:

$$w_N^{n(k+N/2)} = -w_N^{nk} \tag{2.5}$$

Using the periodicity and symmetry of the coefficients w_N^{nk}, investigate how it simplifies the process of DFT operations. Taking $N = 4$ as an example, the matrix form is as follows:

$$\begin{bmatrix} w_4^0 & w_4^0 & w_4^0 & w_4^0 \\ w_4^0 & w_4^1 & w_4^2 & w_4^3 \\ w_4^0 & w_4^2 & w_4^4 & w_4^6 \\ w_4^0 & w_4^3 & w_4^6 & w_4^9 \end{bmatrix} \underset{\Leftrightarrow}{\textit{Periodically}} \begin{bmatrix} w_4^0 & w_4^0 & w_4^0 & w_4^0 \\ w_4^0 & w_4^1 & w_4^2 & w_4^3 \\ w_4^0 & w_4^2 & w_4^0 & w_4^2 \\ w_4^0 & w_4^3 & w_4^2 & w_4^1 \end{bmatrix} \underset{\Leftrightarrow}{\textit{Symmetry}} \begin{bmatrix} w_4^0 & w_4^0 & w_4^0 & w_4^0 \\ w_4^0 & w_4^1 & w_4^0 & w_4^1 \\ w_4^0 & w_4^0 & w_4^0 & w_4^0 \\ w_4^0 & w_4^1 & w_4^0 & w_4^1 \end{bmatrix} \tag{2.6}$$

Matrix operation of DFT:

$$[X(k)] = [x(n)] \; [w_4^{nk}] \tag{2.7}$$

$$\begin{aligned} k=0 \quad X(0) &= x(0)w_4^0 + x(1)w_4^0 + x(2)w_4^0 + x(3)w_4^0 \\ k=1 \quad X(1) &= x(0)w_4^0 + x(1)w_4^1 + x(2)w_4^0 + x(3)w_4^1 \\ k=2 \quad X(2) &= x(0)w_4^0 - x(1)w_4^0 + x(2)w_4^0 - x(3)w_4^0 \\ k=3 \quad X(3) &= x(0)w_4^0 - x(1)w_4^1 - x(2)w_4^0 + x(3)w_4^1 \end{aligned} \tag{2.8}$$

1. Using the periodicity and symmetry of the coefficients can increase the computational speed of the DFT. In the previous example, a DFT requires $N^2 = 16$ multiplication operations, while an FFT requires only 6 multiplication operations.
2. Since the computation of a DFT is proportional to N^2 and N is smaller, a DFT can be performed sequentially on many points of length N. It is decomposed into several small points to operate the DFT. The FFT algorithm is based on the previous two basic ideas to improve the DFT. The FFT algorithm can basically be divided into two categories: extraction by time FFT algorithm, and extraction by frequency FFT algorithm.

2.1.3 FFT decimated by time

To ensure the operational decomposition of the N-point DFT, it is first assumed that N is an integer power of 2, that is, $N = 2^M$ and M is a positive integer.

1. Algorithm principle

$$X(k) = DFT[x(n)] = \sum_{n=0}^{N-1} x(n) w_N^{nk}, \qquad k = 0, 1, \ldots, N-1 \tag{2.9}$$

- Decompose the sequence $x(n)$ into two-point subsequences $N/2$ according to the odd and even numbers of sequence number n:

$$\begin{cases} x(2r) = x_1(r) & \text{even} \\ x(2r+1) = x_2(r) & \text{odd} \end{cases} \qquad r = 0, 1, \ldots, \frac{N}{2} - 1 \tag{2.10}$$

DFT operations are also divided into two groups accordingly:

$$\begin{aligned} x(k) &= DFT[x(n)] = \sum_{n=0}^{N-1} x(n) w_N^{nk} \\ &= \sum_{\text{even}\, n=0}^{N-2} x(n) w_N^{nk} + \sum_{\text{odd}\, n=1}^{N-1} x(n) w_N^{nk} \\ &= \sum_{r=0}^{N/2-1} x(2r)\ W_{\frac{N}{2}}^{rk} + \sum_{r=0}^{N/2-1} x(2r+1)\ w_N^k\ W_{\frac{N}{2}}^{rk} \end{aligned} \tag{2.11}$$

$$X(k) = \sum_{r=0}^{N/2-1} x_1(r) W_{\frac{N}{2}}^{rk} + W_N^k \sum_{r=0}^{N/2-1} x_2(r) W_{\frac{N}{2}}^{rk} = X_1(k) + W_N^k X_2(k) \tag{2.12}$$

Clearly, both $X_1(k)$ and $X_2(k)$ contain DFT for only $N/2$ points, so $X(k)$ also contains DFTs for only $k = 0, 1, \ldots,\ N/2 - 1$ points of the DFT, and the coefficients must also use the periodicity and symmetry of $X(k)$ to find the DFT of the last $N/2$ points of $X(k)$:

$$X_1(k) = \sum_{r=0}^{N/2-1} x_1(r)\ W_{\frac{N}{2}}^{rk} \qquad k = k + \frac{N}{2} \tag{2.13}$$

$$X_1\left(k + \frac{N}{2}\right) = \sum_{r=0}^{N/2-1} x_1(r)\ w_{N/2}^{r(k+N/2)} = \sum_{r=0}^{N/2-1} x_1(r)\ w_{N/2}^{rk} = X_1(k) \tag{2.14}$$

Similarly:

$$X_2\left(k+\frac{N}{2}\right)=X_2(k) \tag{2.15}$$

Using the symmetry of the W factor:

$$W_N^{\left(k+\frac{N}{2}\right)}=-W_N^k \tag{2.16}$$

$$X\left(k+\frac{N}{2}\right)=X_1(k)-W_N^k\ X_2(k), \qquad k=0,1,\ldots,\frac{N}{2}-1 \tag{2.17}$$

The DFT of an N-point is decomposed into the DFT of two $N/2$ points, and the DFT of these two $N/2$ points is then synthesized into an N-point DFT:

$$\begin{cases} X(k)=X_1(k)+W_N^k\ X_2(k) \\ X\left(k+\frac{N}{2}\right)=X_1(k)-W_N^k\ X_2(k) \qquad k=0,1,\ldots,\frac{N}{2}-1 \end{cases} \tag{2.18}$$

The two previous expressions show that computing all sums at $N/2$ points are equivalent to finding the value of $X(k)$ at N points, and it reduces the number of DFT operations by about half.

The operation of the two previous equations can be represented by a special butterfly diagram.

In Figure 2.1, the diagram (a) requires two multiplications and two additions and subtractions. Simplifying Figure 2.1(a) to Figure 2.1(b) requires only one multiplication and two additions and subtractions:

$$\begin{cases} X(k)=X_1(k)+W_N^k\ X_2(k) \\ X\left(k+\frac{N}{2}\right)=X_1(k)-W_N^k\ X_2(k) \end{cases} \tag{2.19}$$

The flow chart of butterfly operations by time extraction is shown in Figure 2.2.

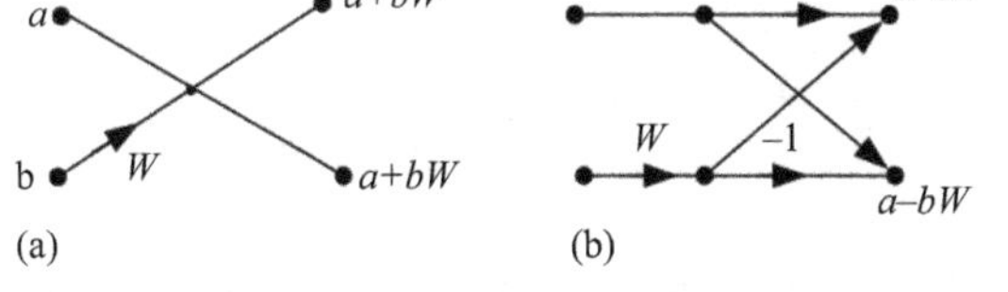

Figure 2.1 Diagram of butterfly signal flow: (a) expression diagram before simplification and (b) expression diagram after simplification

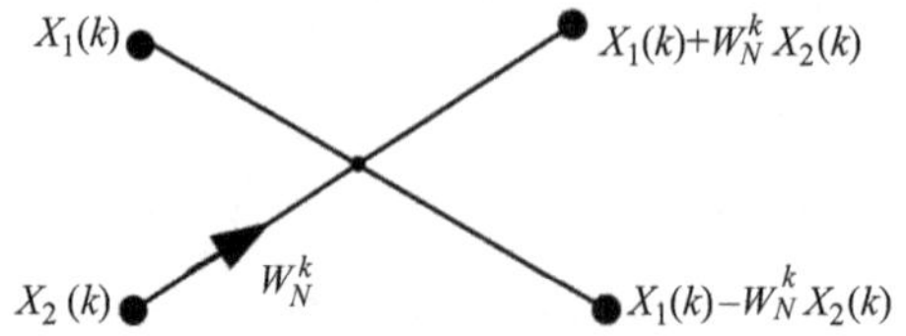

Figure 2.2 Diagram of butterfly signal flow

2.1.4 FFT by frequency extraction

Time-by-time extraction is an alternative FFT algorithm for $N = 2M$.

- The basic idea of the time-based extraction FFT algorithm: (1) $X(n)$ are separated into odd and even levels; (2) $X(k)$ is divided into a level by the first half and the second half.
- The basic idea of the FFT algorithm is based on frequency division: (1) $X(k)$ is separated into odd and even levels; (2) $X(n)$ is divided into two halves by the first and second halves.

Principle of FFT algorithm based on frequency extraction method:

$$X(k) = \sum_{n=0}^{N-1} x(n)\, W_N^{nk} \tag{2.20}$$

Put $x(n)$ in the order of n split in half:

$$\begin{aligned} X(k) &= \sum_{n=0}^{N/2-1} x(n)\, W_N^{nk} + \sum_{n=N/2}^{N-1} x(n)\, W_N^{nk} \\ &= \sum_{n=0}^{N/2-1} x(n)\, W_N^{nk} + \sum_{n=0}^{N/2-1} x\left(n' + \frac{N}{2}\right) W_N^{\left(n'+\frac{N}{2}\right)k} \\ &= \sum_{n=0}^{N/2-1} \left[\, x(n) + W_N^{(N/2)k}\, x(n+N/2) \,\right] W_N^{nk} \end{aligned} \tag{2.21}$$

$$W_N^{N/2} = -1,\quad W_N^{(N/2)k} = (-1)^k = \begin{cases} 1, & k \text{ is even} \\ -1, & k \text{ is odd} \end{cases} \tag{2.22}$$

Based on the odd and even numbers of k, $X(k)$ is further decomposed into two parts:

$$X(k) = \sum_{n=0}^{N/2-1} [x(n) + (-1)^k x(n+N/2)]\, W_N^{nk} \tag{2.23}$$

1. k is an even number, if $k = 2r$

$$\begin{aligned} X(2r) &= \sum_{n=0}^{N/2-1} [x(n) + x(n+N/2)] W_N^{2nr} \\ w_N^{2nr} &= e^{-j\frac{2\pi}{N/2}nr} = w_{N/2}^{nr} \\ &= \sum_{n=0}^{N/2-1} \underbrace{[x(n) + x(n+N/2)]}_{x_1(n)} W_{N/2}^{nr} \\ &= \sum_{n=0}^{N/2-1} x_1(n)\, W_{N/2}^{nr} = DFT[x_1(n)]_{N/2} \end{aligned} \tag{2.24}$$

2. k is an odd number, if $k = 2r + 1$

$$\begin{aligned} X(2r+1) &= \sum_{n=0}^{N/2-1} [x(n) - x(n+N/2)]\, W_N^{n(2r+1)} \\ &= \sum_{n=0}^{N/2-1} \underbrace{[x(n) - x(n+N/2)]\, W_N^n}_{x_2(n)}\, W_{N/2}^{nr} \\ &= \sum_{n=0}^{N/2-1} x_2(n)\, W_{N/2}^{nr} = DFT[x_2(n)]_{N/2} \end{aligned} \tag{2.25}$$

2.2 Hilbert transform

Signal processing plays a fundamental role in a variety of scientific and engineering fields, enabling us to extract valuable information from raw data. Amongst the myriad of signal-processing techniques, the HT stands out as a powerful tool for analyzing signals in the time and frequency domains. This chapter explores the nature of the HT, its mathematical foundations, and its various applications in signal processing. The HT is based on the concept of an analytical signal, which is a complex-valued representation of a real signal. At the heart of the HT is the Hilbert kernel, a mathematical function that acts as a filter to convert the real signal into an analytical signal. The Hilbert kernel is convolved with the original signal to obtain the real part of the signal and the imaginary part of the HT.

The HT provides valuable insight into the amplitude and phase characteristics of a signal. Applying the HT to a real-valued signal, we can obtain the corresponding analytic signal, which consists of the real part of the original signal and the HT as the imaginary part. This analytical representation allows us to extract instantaneous amplitude and phase information, which allows us to accurately analyze the envelope, frequency content, and time-varying characteristics of the signal.

A prominent application of the HT is envelope detection. By calculating the amplitude of the analyzed signal obtained through the HT, we can accurately extract the envelope of the original signal. This facilitates the identification of signal variations, amplitude modulation, and transient events, allowing us to effectively detect anomalous or significant features of various signals, such as speech, music, and vibration data. The HT plays a key role in frequency analysis. By applying the HT to a real signal, we can obtain an analytical representation of it from which instantaneous phase and amplitude can be derived. This allows the instantaneous frequency of the signal to be accurately estimated using methods such as instantaneous frequency calculations and the Hilbert–yellow transform. The HT also has applications in demodulation tasks, where it is possible to separate the carrier signal from the modulated signal. The HT has applications in a variety of fields. In audio and speech processing, it helps in pitch detection, speech analysis, and music processing. In vibration analysis, it helps in fault diagnosis, modal analysis, and condition monitoring of mechanical systems. In addition, the HT has proved useful in communication systems, biomedical signal processing, seismic analysis, and image processing.

The HT is a powerful technique in signal processing, enabling us to unravel the complex dynamics of signals in both the time and frequency domains. Its ability to extract instantaneous amplitude, phase, and envelope information provides valuable insights into signal characterization, facilitating work such as envelope detection, frequency analysis, and modulation analysis. As signal processing continues to evolve, the HT remains an invaluable tool for researchers and practitioners in a variety of fields, aiding the exploration and interpretation of complex signals. By harnessing the analytical power of the HT, we can gain greater insight into the hidden complexity of signals, paving the way for advances in fields ranging from telecommunications and biomedical engineering to audio processing and structural analysis.

Function $u(t)$ of the HT is defined as

$$H(u)(t) = \frac{1}{\pi} p \cdot v \cdot \int_{-\infty}^{+\infty} \frac{u(\tau)}{t - \tau} d\tau \tag{2.26}$$

where "$p \cdot v \cdot$" stands for the Cauchy principal value, which is the integral value of the flawed integral. "$p \cdot v \cdot$" can also be replaced by p.

From the previous definition, the HT of a function u of the HT of a function is the convolution of that function and the convolution of $h = 1/\pi t$:

$$H(u)(t) = u \times h = \frac{1}{\pi} p \cdot v \cdot \int_{-\infty}^{+\infty} \frac{u(\tau)}{t - \tau} d\tau \tag{2.27}$$

2.3 Short-time Fourier transform

STFT is an analytical tool that decomposes a signal into two dimensions, namely, time and frequency. It has a wide range of applications in the field of signal processing. The STFT decomposes a signal into two dimensions, time, and frequency, by window sampling the signal in the time domain and performing a Fourier transform in the frequency domain. This allows us to analyze the spectral information of the signal at different times and frequencies. Window functions are used in STFT to control the truncation of the signal in the time domain. Common window functions include the Hanning window and the rectangular window, which reduce the problem of spectral leakage. STFT provides the ability to resolve signal characteristics in time and frequency. Shorter window lengths provide better temporal resolution but poorer frequency resolution, while longer window lengths provide better frequency resolution but poorer temporal resolution. Doing DFT on a nonstationary signal does not give temporal information about the changes in the signal. For example, many signals appear and then disappear over a while. DFT cannot determine the order in which the different signals appear, whereas STFT takes out the signals for each period. After a short windowing period, the DFT comes to reflect the change in signal over time.

Definition: STFT is also called short-time spectrum (by adding a window):

$$X_n\left(e^{jw}\right) = \sum\nolimits_{m=-\infty}^{\infty} x(m)w(n-m)e^{-jwm} \tag{2.28}$$

Characteristics of the short-time spectrum:

1. Time-varying: It is a function of angular frequency ω and time n.
2. Periodicity: It is a periodic function concerning ω with a period of 2π.

The STFT is the standard Fourier transform of a speech signal with a window. It is distinguished from the standard Fourier transform by the subscript $n{\cdot}\omega(n-m)$ is a sequence of window functions. Different sequences of window functions will give different Fourier transform results. The STFT has two independent variables: n and ω, so it is both a discrete function of time n and a continuous function of angular frequency ω. In the same way, as for the DFT and the continuous Fourier transform, if $\omega = 2\pi k/N$, a discrete STFT is obtained, which is effectively sampled in the frequency domain:

$$Xn\left(e^{j\frac{2k\pi}{N}}\right) - Xn(k) - \sum_{m=-8}^{\infty} x(m)w(n-m)e^{-j\frac{2k\pi m}{N}}, \qquad 0 \le k \le N-1 \tag{2.29}$$

Both formulas have two interpretations:

1. When n is fixed, they are the sequence $\omega(n-m)x(m)$ of the standard Fourier transform or the standard discrete Fourier transform $-\infty < m < \infty$. At this point, it has the same properties as the standard Fourier transform and $X_n(k)$ has the same properties as the standard discrete Fourier transform.
2. When ω or k is fixed, and $X_n(k)$ are regarded as functions of time n. They are the convolution of a sequence of signals and a sequence of window functions, at which point the window acts as the equivalent of a filter.

Relationship between frequency resolution Δf, sampling period T and addition window width N:

$$\Delta f = \frac{1}{NT} \tag{2.30}$$

Effect of window shape on STFT:

1. Rectangular window—narrow main flap, slow decay
2. Hamming window—wide main flap, fast decay

The effect of window width on the short-time spectrum:

1. Long window width—high-frequency resolution
2. Short window width—low-frequency resolution

Based on the definition of the power spectrum, the relationship between the short-time power spectrum and the STFT can be written as

$$S_n(e^{j\omega}) = X_n(e^{j\omega}) \cdot X_n^*(e^{j\omega}) = |(e^{j\omega})|^2 \tag{2.31}$$

where "*" represents the complex-conjugate operation. At the same time, the power spectrum is the Fourier transform of the short-time autocorrelation function:

$$R_n(K) = \sum_{m=-8}^{\infty} w(n-m)x(m)w(n-k-m)x(m+k) \tag{2.32}$$

The STFT is written in an alternative form later. The standard Fourier transforms of the signal sequence and the window sequence are shown in the following equation:

$$X(e^{j\omega}) = \sum_{m=-\infty}^{\infty} x(m)e^{-j\omega m} \tag{2.33}$$

When n takes a fixed value, the Fourier transform of $\omega(n-m)$ is shown as follows:

$$\sum_{m=-\infty}^{\infty} w(n-m)e^{-j\omega m} = e^{-j\omega m} \cdot W(e-^{j\omega}) \tag{2.34}$$

According to the frequency-domain convolution theorem of the Fourier transform, there is

$$Xn(e^{j\omega}) = X(e^{j\omega}) \times [e^{-j\omega m} \cdot W(e^{-j\omega})] \tag{2.35}$$

Multiplying the waveform by the window function not only allows the waveform to be slowly reduced to zero without causing sharp changes at either end of the window but also corresponds to convolving the signal spectrum with the Fourier transform of the window function.

The window function should have the following characteristics:

1. High-frequency resolution, that is, narrow, and sharp main flaps (rectangular windows).
2. Small spectral leakage through the convolution in the other frequency components, that is, high side lobe attenuation (Hamming window).

Window width $\downarrow\rightarrow$ frequency resolution $\uparrow$ time resolution $\downarrow$
Window width $\downarrow$ $\rightarrow$ frequency resolution $\downarrow$ time resolution $\uparrow$

2.4 Discrete wavelet transform

The DWT is a signal analysis tool that provides localization information in terms of time and frequency. Compared to the Fourier transform and STFT, the DWT has better time–frequency localization properties and provides finer frequency and time resolution. DWT uses wavelet functions of different scales and shifts to decompose and reconstruct the signal. The scale represents the frequency of the wavelet function, while the shift represents the position of the wavelet function in time. The DWT decomposition process is achieved by successively decomposing the signal in the low-frequency band (approximation) and the high-frequency band (detail). At each scale, the signal is divided into two parts: approximate and detailed. The approximate part reflects the low-frequency content of the signal at that scale, while the detailed part reflects the high-frequency details of the signal at that scale. In the decomposition process, the signal is convolved by a low- and a high-pass filter to obtain the approximate and detailed parts. The approximate part is then downsampled to reduce the sample rate, while the detailed part is kept at the original sample rate. In this way, the signal information at each scale is decomposed into approximate and detailed parts in different frequency bands. The DWT reconstruction process is achieved by filtering and upsampling the approximate and detailed parts. By inverse filtering and upsampling, the approximate and detailed parts at each scale can be reconstructed as an approximate reconstruction of the original signal. The advantage of DWT is that it provides local information, allowing for a more refined analysis of the signal. It has a wide range of applications in areas such as signal processing and data compression, including image processing, audio processing, video compression, signal noise reduction, feature extraction, and so on. It is important to note that DWT is a multi-scale analysis method and, unlike the Fourier transform and STFT, it has localization properties in both the time and frequency domains. This makes DWT more applicable in some applications than other transforms but also requires appropriate wavelet function selection and parameter tuning to suit different signal characteristics.

2.4.1 Wavelet function scale discretization methods

For continuous wavelets, the scale a, the time t and the time-dependent offset τ are all continuous. If a computer is used to calculate this, it must be discretized to obtain the DWT.

1. Discretization of scales: A power-level discretization of a scale.

$$a = a_0^j, a_0 > 0, j \in Z \tag{2.36}$$

The corresponding wavelet function is shown as follows:

$$a_0^{\frac{-j}{2}} \psi[a_0^{-j}(t - \tau)], \quad j = 0, 1, 2, \ldots \tag{2.37}$$

The value of the base a_0 of the power series reflects the degree of discretization of the scale.

2. Displacement discretization:

$$\tau = ka_0^j \tau_0 \tag{2.38}$$

Usually, a uniform discrete value is taken for τ to cover the entire time axis, and τ satisfies Nyquist's sampling theorem. At $a = 2j$, the response sampling interval along the τ axis is $2^j \tau_0$. In the case of $a_0 = 2$, an increase of 1 in j doubles the scale a and reduces the corresponding frequency by half. At this point, the sampling rate can be reduced by half without causing a loss of information to be induced.

Thus, at scale j, the sampling interval can be expanded by c without causing a loss of information since the width of $\varphi(a_0^{-j} t)$ is a_0^j times that of $\varphi(t)$.

Here, $\varphi_{a,\tau}(t)$ can be written as

$$a_0^{-\frac{j}{2}} \psi[a_0^{-j}(t - ka_0^j \tau_0)] = a_0^{-\frac{j}{2}} \psi[a_0^{-j} t - k\tau_0)] \tag{2.39}$$

The DWT is defined as

$$WT_f(a_0^j, k\tau_0) = \int f(t) \psi^*_{a_0^j, k\tau_0}(t) dt, j = 0, 1, 2, \ldots, \qquad k \in Z \tag{2.40}$$

In general, $\tau_0 = 1$, the expression was shown as follows:

$$\psi_{j,k}(t) = 2^{-\frac{j}{2}} \psi(2^{-j} t - k) \tag{2.41}$$

At this point, the corresponding WT_f is shown as follows:

$$WT_f(j, k) = \int f(t) \psi^*_{j,k}(t) dt \tag{2.42}$$

Two issues in the discretization process

1. Whether the discrete wavelet can completely characterize the full information of the function $f(t)$.
2. Whether any function $f(t)$ can be expressed as a weighted sum in units of $\varphi_{j,k}(t)$.

2.5 Empirical mode decomposition

EMD is a data-adaptive signal decomposition method for decomposing nonlinear and non-smooth signals into several eigenmode functions (referred to as IMFs).

EMD was proposed by Hongjun Huang in 1998. It does not rely on any predetermined basis functions but performs adaptive decomposition based on the characteristics of the signal itself.

First, the upper and lower envelopes are found from the original signal. The upper envelope is a continuous curve connecting the local maxima of the signal and the lower envelope is a continuous curve connecting the local minima of the signal. By smoothing the upper and lower envelopes, the local average curve of the signal is obtained. The local mean curve is obtained by summing the upper and lower envelopes and dividing by 2. The local mean curve is subtracted from the signal to obtain the local mode of vibration of the signal. The local vibration modes reflect the vibration characteristics of the signal at different scales and frequencies. The extracted local vibration modes are examined and used as the final eigenmode function (IMF) if they satisfy the following two conditions. The advantage of EMD is its ability to accommodate the decomposition of nonlinear and nonstationary signals into a series of IMFs with different frequencies and amplitude characteristics. This allows EMD to be widely used in signal processing, vibration analysis, image processing, and other areas such as signal denoising, time–frequency analysis, pattern recognition, and so on. It should be noted that EMD may have some problems in practical applications, such as mode conflation, boundary effects, and noise sensitivity. To overcome these problems, some improved EMD methods, such as fast EMD and sifting EMD, have been proposed and widely studied.

The starting point of EMD is to consider the oscillations within the signal as local. If we want to look at the variation between the two adjacent extremes of the evaluated signal $x(t)$ (two extrema, at $t-$ and $t+$), we need to define a high-frequency component $\{d(t),\ t- \leq t \leq t+\}$ which corresponds to the oscillation, which is between the two extrema, through the extrema. To complete the diagram, we also need to define a low-frequency component $m(t)$ such that $(x(t) = m(t) + d(t), t- \leq t \leq t+)$. For all vibrational components of the whole signal, if we can find a suitable method for such a decomposition, this process can be applied to all remaining components of the local trend so that the constituent components of the signal can be abstracted by iteration.

For a given signal $x(t)$, the following steps are performed to perform an efficient EMD decomposition:

1. Find all extreme points of the desired $x(t)$.
2. Form the lower envelope $e_{\min}(t)$ for the very small values and the upper envelope $e_{\max}(t)$ for the very large values by interpolation.
3. Calculate the mean value $m(t) = (e_{\min}(t) + e_{\max}(t))/2$.
4. Extract the details $d(t) = x(t) - m(t)$.
5. Repeat the appeal steps for the residual $m(t)$.

In practice, the previous process needs to be redefined by a screening process, where the first iteration step of the screening process is to repeat from steps 1 to 4 for the detail signal $d(t)$ until the mean value of $d(t)$ is 0, or some stopping criterion is satisfied before stopping the iteration.

Once the stopping criterion is satisfied, the detail signal $d(t)$ at this point is called the IMF, and the $d(t)$ corresponding to the residual signal is calculated using step 5. Through the previous process, the number of extremal points is accompanied by the generation of fewer and fewer residual signals, and the whole decomposition process produces a finite number of modal functions (IMFs). Modal functions and residual signals can be analyzed spectrally, but this spectral analysis should not be viewed from a narrow perspective. First, it should be emphasized that even in the case of resonant oscillations, the high and low frequencies generated by applying the previous method are only localized, and there is no way to generate a predefined band filter (e.g., WT) for identification. The chosen mode function corresponds to an adaptive (signal-dependent) time-varying filter. The EMD algorithm relies on a range of options that require user control and expertise. Our aim here is to find more accurate options and to make improvements to the original algorithm.

2.5.1 The sampling rate, interpolation methods, and edge effects

The basic operation of EMD is to estimate the interpolation curve between the upper and lower envelopes as the extreme value points. The performance of the chosen interpolation method is a very important parameter. Our experiments are identifying cubic spline interpolation methods as the preferred choice. Other interpolation methods (linear or polynomial) increase the number of screening iterations and produce over-decomposed signals that are scattered around the modal function. Second, since this algorithm applies to real discrete-time signals, care must be taken that the extrema are correctly identified, a necessary condition that requires extensive oversampling (this will be studied further in Section 2.5.4) Finally, boundary states must be considered, as very small errors can spread over a finite length of measurement data. As a consideration, better results can be obtained by mirroring the extrema along the terrain by adding them near the boundary.

2.5.2 Stopping criteria for screening

When the screening process is complete, the extracted mode function will be satisfactory. There are two necessary conditions for this: First, the number of extreme points and the number of over-zero points must differ by at most 1, and second, the average value of the upper and lower envelopes must be approximately equal to zero or satisfy some criterion. The size of the mean needs to be calculated relative to the size of the corresponding modal function, but terminating the iterative process with a threshold that is too low can lead to the problems described earlier (over-iterations produce over-decomposition).

2.5.3 Local EMD

In classical EMD applications, the screening iterations fit over the entire length of the data and are sought to be small enough as long as there are local regions present (appearing at the envelope mean). However, as already mentioned in the text, because a good local approximation has the disadvantage of data contamination from the rest of the signal, the whole signal can become over-iterated. This is particularly likely to happen with over-decomposition of the same amplitude and spread to other neighboring mode functions. The hierarchical and nonlinear nature of the original algorithm does not guarantee that the EMD of a sequential signal will be a sequence of every EMD.

2.5.4 Online EMD

The second improved algorithm is based on the fact that the screening process relies on the interpolation of extreme value points, so that for a given data length, only a limited number of extreme value points need to be used (three spline interpolations of five minima and five maxima). Thus, the modal function extraction does not require knowledge of the entire signal (or the previous residual signal). The necessary conditions for extracting modal functions also apply to the filtering process, which prevents possible discontinuities, so that the whole signal needs to be known, the number of filtering runs is consistent with a priori knowledge, and a small number of iterations (less than 10, especially 4) can produce enough meaningful IMFs. To apply the online EMD algorithm more effectively, we propose to add smoothing to the local algorithm window. New valid data enter the leading edge of the window, and when the stopping criterion is met, the entire block of data enters the trailing edge. Based on such a stop, the IMF and its corresponding residual component can thus be calculated. The entire algorithm can therefore be applied to the residual components and subsequently abstracted from the remaining modal functions to the valid components.

2.5.5 Performance basis

Since EMD is essentially defined as an algorithm, but not considered as an analytical method, its performance estimation is difficult and requires extensive simulation experiments. We will report on his principles in two ways: one from the extraordinary properties of EMD as demonstrated by previous applications to real signals, and the other from an understanding of decomposition.

2.5.6 Components and sampling rates

When we analyze a single-component signal, the EMD, as a resolving unit, ideally separates only one mode function (which accurately identifies the single component) and does not contain any residual components. However, it is difficult to obtain such an idealized decomposition even when the effects of edge are removed, as the effect of the sampling rate on the EMD decomposition cannot be avoided. If few points are sampled in a cycle, then the extreme points sampled at such a sampling rate will not be accurate.

2.6 Principal-component analysis

PCA is a commonly used technique for multivariate data reduction and feature extraction. It uses a linear transformation to project the original data onto a new coordinate system so that the projected data have the maximum variance for data reduction and feature extraction. First, the original data are normalized so that each feature has a mean of zero and a variance of units. This is to ensure that each feature has the same importance in subsequent calculations. The linear correlation between the data features is described by calculating a covariance matrix for the normalized data. The elements of the covariance matrix represent the degree of correlation among the different features. The covariance matrix is decomposed into eigenvalues to obtain the eigenvalues and corresponding eigenvectors. The eigenvalues represent the variance of the data in the direction of the corresponding eigenvectors. The eigenvector with the largest eigenvalue is selected as the principal component based on its magnitude. The principal component is a linear combination of the original eigenvectors and represents the main pattern of variation in the data.

The original data are projected onto the selected principal components to obtain reduced dimensional data. Projected data can be thought of as the projection coefficients of the original data in the direction of the principal components.

2.6.1 Advantages of PCA

Data dimensionality reduction: By selecting the most important principal components, high-dimensional data can be reduced to lower dimensions, thus reducing data storage and computational costs. PCA can help to identify and extract key features from the data, thus helping to understand the structure and patterns of the data. Reduced dimensional data can be visualized in a lower dimensional space, thus facilitating the observation and analysis of the data. It is important to note that PCA assumes that the data obey a linear model, with principal components chosen based on the magnitude of the variance. In some cases, nonlinear relationships and other dimensionality reduction methods (e.g., stream learning) may be more applicable. In addition, PCA is sensitive to the distribution and noise of the data, so the characteristics of the data and preprocessing methods need to be considered when applying PCA.

In terms of the physiological characteristics of the human brain, the human brain perceives the outside world through various means, resulting in the human acquiring too many dimensions of information. If the human brain does not downscale the acquired information, then the efficiency and accuracy of the human brain in processing the information will be reduced, so the human brain goes through a complex downscaling process for all these perceptual neural processes. The PCA method is widely used in data analysis from neuroscience to computer graphics. This is because it is a simple, nonparametric method for extracting data-relevant information from complex data sets. Our motivation for performing PCA is the desire to compute the most important bases in the noisy data space to reexpress this data space. However, these new bases are often hidden in complex data structures, and we need to filter out the noise and find new bases for reconstructing the data space.

The PCA method is a highly versatile approach with the advantage of being able to reduce the dimensionality of the data. We use the PCA method to find the main elements of the dataset, select the most important parts, and omit the rest of the dimensions, thus achieving a reduction and simplification of the model, indirectly compressing the data while retaining a great deal of information about the original data, just like the dimensionality reduction process performed by the human brain during perceptual neural processing. This is why PCA methods are widely used in the fields of machine learning and pattern recognition as well as computer vision. In face recognition, assume that the training set is 30 different $N \times N-$ sized face images. Considering that each pixel in an image is one-dimensional information, then a pair of images is a two-dimensional vector. This is because there is a great deal of similarity in the structure of faces, and even greater similarity if they are the same face. And what we want is to be able to represent faces in terms of faces, not pixels. We can then use the PCA method on the 30 training set images to find the similarity dimensions in these images. After we have extracted the most important principal components, we let the recognized images be compared to the original images with the principal-component dimensions changed to measure the similarity of the two images.

In terms of image compression, we can also perform image compression by the PCA method, also known as Hotelling or Karhunen and Leove transform. By extracting the principal components of the image, removing some minor components, and then transforming back to the original image space, the image is compressed to a large extent due to the dimensionality reduction, while the image still retains much of the important information of the original image. The PCA method is a process of mapping the data space to a low-dimensional subspace using an orthogonal transformation. The corresponding set of basis vectors should satisfy orthogonality, and the state subspace consisting of the set of basis vectors should ideally take into account the correlation of the data. After the transformation of the space of the original data set, the correlation between individual data samples should be reduced to a minimum.

To summarize the steps of the PCA algorithm:

1. Assume the size of the data is $m \times n$,
   1. Form the original data into a matrix X of n rows and m columns.
   2. Zero-mean for each row of X, that is, subtract the mean of that row.
   3. Derive the covariance matrix $C = \frac{1}{m}XX^T$.
   4. Find the eigenvalues of the covariance matrix and the corresponding eigenvectors.
   5. Arrange the eigenvectors into a matrix in rows from top to bottom according to their corresponding eigenvalues, and take the rows k to form the matrix P.
   6. $Y = PX$ is the data after dimensionality reduction to k dimensions.

2.7 Singular-value decomposition

SVD is an important matrix decomposition method that decomposes a matrix into the product of three matrices. SVD has a wide range of applications in many fields such as data compression, signal processing, image processing, recommendation systems, and so on.

1. Eigenvalues:

   If a vector v is an eigenvector of a square matrix A, it will necessarily be expressed in the following form:

$$Av = \lambda v \tag{2.43}$$

   where λ is referred to as the eigenvalue corresponding to the eigenvector v. The eigenvectors of a set of matrices are a set of orthogonal vectors. Eigenvalue decomposition is the decomposition of a matrix into the form as follows:

$$A = Q\Sigma Q^{-1} \tag{2.44}$$

   where Q is a matrix consisting of the eigenvectors of this matrix A. Each element on the diagonal is an eigenvalue. It must be clear that a matrix is a linear transformation since the vector obtained by multiplying a matrix by a vector is equivalent to a linear transformation of that vector.

   When the matrix is high-dimensional, then this matrix is a linear variation of the high-dimensional space. This linear change may not be represented by a picture, but it is conceivable that this transformation also has many transformation directions. The first N eigenvectors obtained by eigenvalue decomposition correspond to the most important N directions of change of this matrix. We can approximate this matrix (transformation) by using the first N directions of change. In summary, the eigenvalue decomposition gives us the eigenvalues and the eigenvectors. Each eigenvector can be understood as a linear subspace, which can be used for many things. However, there are many limitations to the eigenvalue decomposition, for example, the transformed matrix must be square.
2. Singular values:

In the real world, most matrices we encounter are not square matrices. For example, if there are N students, each with M grades, the resulting matrix of size $N \times M$ is not square. SVD is a decomposition method that can be applied to any matrix. Suppose A is an $N \times M$ matrix, then the resulting U is an $N \times N$ square matrix (the vectors inside are orthogonal, and the vectors inside U are called left singular vectors). Σ is an $N \times M$ matrix (all elements except the diagonal are 0, and the elements on the diagonal are called singular values). V^T (the transpose of V) is an $M \times M$ matrix, and the vectors inside are also orthogonal (the vectors inside V are called right singular vectors). The sizes of the several multiplied matrices can be reflected by the following equation:

$$A_{m\times n} = U_{m\times m} \times \Sigma_{m\times n} \times V_{n\times m}^{T} \tag{2.45}$$

Singular values and eigenvalues correspond to each other. Taking the transpose A^T of matrix A to get a square matrix to find the eigenvalues:

$$\left(A^T A\right) V_i = \lambda_i v_i \tag{2.46}$$

The v obtained is the right singular vector that we have before:

$$\sigma_i = \sqrt{\lambda_i} \tag{2.47}$$

$$u_i = \frac{1}{\sigma_i} A v_i \tag{2.48}$$

where σ is the singular value as described earlier and u is the left singular vector as described earlier. The singular values σ are similar to the eigenvalues in that they are also ordered from largest to smallest in the matrix Σ.

PCA dimensionality reduction involves finding the top d eigenvectors of the sample covariance matrix $X^T X$ and projecting the data onto the matrix spanned by these eigenvectors for low-dimensional projection. Computing the covariance matrix $X^T X$ is a computationally intensive task, especially when the number of samples and features is large. It is worth noting that SVD can also be used to obtain the matrix of the top d eigenvectors of the covariance matrix $X^T X$. However, one advantage of SVD is that certain implementations can directly obtain the right singular matrix V without explicitly calculating the covariance matrix $X^T X$. This means that the PCA algorithm can be implemented using SVD without the need for eigen decomposition. This approach works particularly well when dealing with a large number of samples. The implementation of the PCA algorithm in scikit-learn uses SVD instead of the conventional eigen decomposition.

Chapter 3

AI-based electric motor fault diagnosis for electric powertrain in EVs

As the crucial step in fault detection and diagnosis (FDD), accurate and efficient classification performance is extremely important. As an attractive method, machine learning (ML) includes a variety of model types: (1) supervised learning, which is responsible for analyzing training data and generating inference functions through algorithms; (2) unsupervised learning to draw conclusions from unlabeled data. The most common unsupervised learning method is cluster analysis, which is mainly used to discover hidden patterns in grouped data. Deep learning (DL) is a sub-domain of ML, and its related algorithms are inspired by the structure and function of the brain (i.e., artificial neural networks (ANNs)). DL uses a hierarchical framework of non-nearline transformations of input information to construct statistical models and output outcomes. Traditional ML is capable of adapting to a wide range of data quantities, particularly in circumstances with small data volumes. On the other hand, if the amount of data increases rapidly, the effect of DL will be more prominent. Figure 3.1 illustrates the flowchart of traditional ML and DL. The application of classic ML and popular DL algorithms in motor FDD are introduced in this section.

3.1 Support vector machine (SVM)-based approach

A support vector machine (SVM) is a binary classification model that attempts to build a line between two classes of points by mapping the feature vector of an instance to certain points in space. SVM is applied in regression analysis for data classification and system parameter estimation.

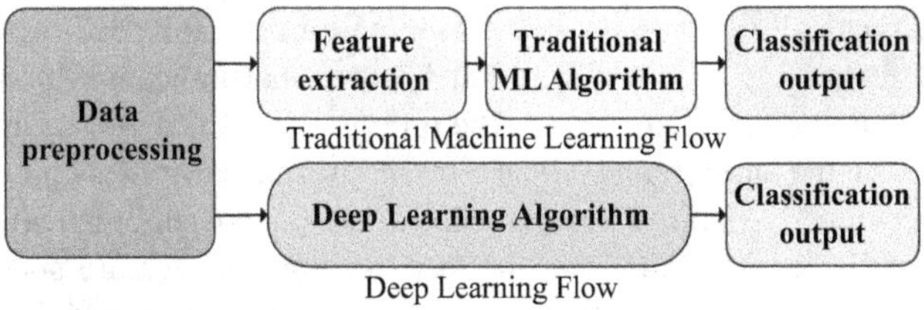

Figure 3.1 Flowchart of traditional ML and deep learning

SVM has various applications such as handwriting recognition, image recognition, and others. SVM is particularly useful for small samples and database instances. Because of its appealing characteristics and strong analytical results, SVM is becoming increasingly popular in the field of motor drive system fault diagnosis. Figure 3.6 illustrates the optimal hyperplane for binary classification by SVM:

$$\min \Phi(w) = \frac{1}{2}\|w\|^2 \tag{3.1}$$

$$\text{subject to } y_i(w \cdot x_i + b) \geq 1; \qquad i = 1, 2, \ldots, n \tag{3.2}$$

where w stands for the weight vector, which describes the hyperplane, x_i stands for the input vector, y_i stands for the label associated with y_i, and b stands for a scalar threshold. For the case of nonlinear separation, in high-dimensional spaces, the SVM can construct a hyperplane that allows for linear separation. Only the transformation $\Phi(x)$ from N-dimensional space to Q-dimensional function space of the data is used in SVMs. Since the inner product can be replaced, the mapping translation can be achieved with a kernel operation. By retaining the influence of high-dimensional transformations, it reduces the computational load. This is the kernel function: $k(x, x_i) = \Phi(x).\Phi(x_i)$. Many functions can be used as the kernel function. The Gaussian radial basis function (RBF) is a well-known kernel, the equation is as follows:

$$k(x, x_i) = \exp \exp\left(-\gamma\|x - x_i\|^2\right); \qquad \gamma = \frac{1}{2\sigma_b^2} > 0 \tag{3.3}$$

After optimization, the basic SVM structure can be described as

$$f(x) = \sum\nolimits_{i=1}^{m} \{y_i a_i k(x, x_i)\} + b \tag{3.4}$$

SVM has a nonlinear multi-label classification function. For multiple fault diagnosis, an intelligent fault detection system based on a multivariate integrated incremental support vector machine (MEISVM) is presented in [63] for multi-fault diagnosis. This method can detect a variety of faults, including complex compound faults and faults with different severity thresholds. Reference [64] proposed a grid search support vector machine (GSSVM) approach based on redefined dimensionless indicators (RDI) extracted maximum correlation and minimum redundancy feature selection. Minimum redundancy maximum relevance (mRMR) was used to automatically eliminate redundant and irrelevant features from high-dimensional features to obtain more fault features that indicated the useful information hidden in the vibration signal. The proposed RDI as a new fault feature can effectively solve the shortcomings of the traditional dimensionless index and has a stronger discriminative ability for mechanical faults and automatically eliminate redundant and irrelevant features in the high-dimensional feature space. The correlation between features and output class labels is maximized, and the redundancy between features is minimized. However, the dimensionless metric is sensitive to faults rather than to operating conditions, whether the classification effect is significant when the operating conditions of the motor change are not given as a validation. A novel intelligent fault detection approach for rolling bearings was proposed based on composite multiscale fuzzy entropy (CMFE) and ensemble support vector machines (ESVMs) to extract the nonlinear features that were embedded in the vibration signal [65]. CMFE was used to extract hidden nonlinear fault characteristics from rolling bearing vibration signals, and then ESVM was used to construct a multi-fault classifier to accomplish an automated intelligent diagnosis of rolling bearings. In [66], several investigations were carried out to record current conditions during various motor power supply activities, such as internal stator winding short-circuit failures and supply voltage imbalance at different load speeds. A recursive feature elimination algorithm based on the support vector machine (SVM-RFE) was applied to select and maximize the number of appropriate features to be used for classification. It was worth noting that two different feature sets were created. One of them contained load-level details, which were intentionally hidden in the other one. The selected features were used to identify various stator-winding fault conditions using a support vector regression (SVR)-based classifier. It has been found that the performance of the classifier was better when the load level information was included in the functional layer rather than the load level information was hidden. To make the fault classification algorithm uninterrupted at different load levels, two additional functions were extracted from the Park vector modulus using detrended fluctuation analysis (DFA). Ultimately, the SVR-based classifier was found to be accurate in detecting and classifying internal motor faults. A satisfactory response was shown in distinguishing between supply voltage imbalance and internal stator faults. Full-spectrum cascade analysis of rotating machine vibrations is an effective method. Abrupt changes in acceleration signals of rotor faults can be detected and the characteristic spectrum of faults is displayed in a full-spectrum cascade diagram. In addition, the fault diagnosis problem becomes more confusing when the current characteristics due to power supply voltage imbalance are significantly similar to those due to internal stator-winding faults. It is important to select, rank, and optimize the number of valid features used for classification. RFE uses some mathematical or heuristic rules to train the SVM classifier with features that minimize the margins. This feature elimination process is repeated until some stopping criterion is satisfied. Removing features during the iterative process simplifies the computation, but this may lead to suboptimal solutions. The full spectrum experimental data were used for SVM training and excellent classification results were obtained [67].

Traditional statistical pattern recognition methods only have theoretical performance guarantees when the number of samples tends to infinity. Statistical learning theory (SLT) investigates ML problems under the finite sample scenario, and SVMs have their theoretical foundation in SLT. Traditional statistical pattern recognition methods emphasize the minimization of empirical risk while performing ML. Generality refers to the ability of a learning machine to correctly predict future outputs. In some cases, when the training error is too small, it can actually lead to a decrease in generalization ability. For example, given a set of training samples (x, y), where x is distributed over a range of real numbers and y takes values between [0,1], we can always fit them using $y = \sin\ (w \times x)$, resulting in zero training error. According to SLT, the actual risk of a learning machine consists of two components: empirical risk and confidence interval. Learning methods based on the principle of empirical risk minimization only emphasize the minimization of empirical risk on training samples, without minimizing the confidence interval value. Therefore, their generalization ability is poor. Vapnik proposed SVM, which constrains the optimization problem based on the training error and aims to minimize the confidence interval value. SVM is a learning method based on the principle of structural risk minimization, and its generalization ability is significantly superior to some traditional learning methods. The years from 1992 to 1995 marked its growth period. The solution of SVM is globally unique and optimal because it ultimately transforms into a quadratic programming problem. SVM has demonstrated many unique advantages in solving small-sample, nonlinear, and high-dimensional pattern recognition problems, and it can also be extended to other ML problems such as function approximation. Joachims recently used SVM for text classification in Reuters-21578 and claimed that it outperformed all other methods published to date.

Linear discriminant functions and discriminant surfaces: A linear discriminant function is a function consisting of linear combinations of the components of x:

$$g(x) = w^T x + w_0 \tag{3.5}$$

Two types of cases: The decision rules for the two types of problems are as follows:

If $g(x) > 0$, then decide that x belongs to $C1$.

If $g(x) < 0$, then it is decided that x belongs to $C2$.

If $g(x) = 0$, then x can be arbitrary to a class or reject the decision.

Figures 3.2 and 3.3 represent a simple linear classifier with d input units, each unit corresponding to a component value of the input vector in each dimension.

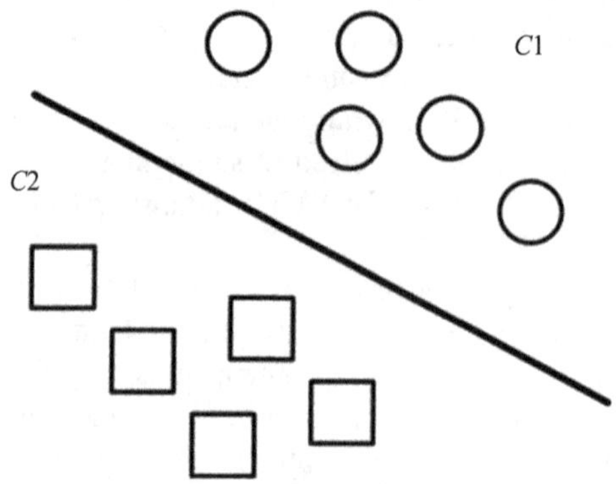

Figure 3.2 Discriminant surface

The equation $g(x) = 0$ defines a decision plane that separates points classified as $C1$ from points classified as $C2$. When $g(x)$ is a linear function, the plane is called a "hyperplane." When both $X1$ and $X2$ are on the decision surface, it indicates that w is orthogonal to any vector on the hyperplane, and w is the normal vector of the hyperplane.

There are multiple methods for designing a multi-class classifier using linear discriminant functions. For example, a k-class problem can be transformed into k binary problems, where the ith problem involves using a linear discriminant function to separate points belonging to class Ci from points not belonging to class Ci. A more complex method involves using $k(k-1)/2$ linear discriminant functions to divide the samples into k classes, with each linear discriminant function classifying only two classes among them.

3.1.1 Generalized linear discriminant function

In one-dimensional (1-D) space (Figure 3.4), there is no linear function that can solve the following division problem (where black and red represent two different classes of data). It indicates that linear discriminant functions have certain limitations.

If we establish a quadratic discriminant function $g(x) = (x-a)(x-b)$, we can effectively solve the classification problem mentioned earlier (Figure 3.5). The decision rule remains the same: If $g(x) > 0$, then x is classified as belonging to $C1$; if $g(x) < 0$, then x is classified as belonging to $C2$; if $g(x) = 0$, then x can be assigned to either class or can be rejected.

Quadratic discriminant function $(x) = (x-a)(x-b)$. It can be written in the following general form:

$$g(x) = c_0 + c_1 x + c_2 x^2 \tag{3.6}$$

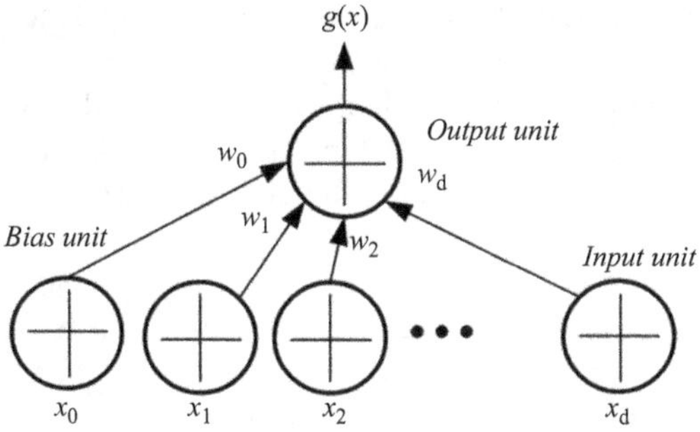

Figure 3.3 Diagram of a linear classifier

Figure 3.4 One-dimensional space

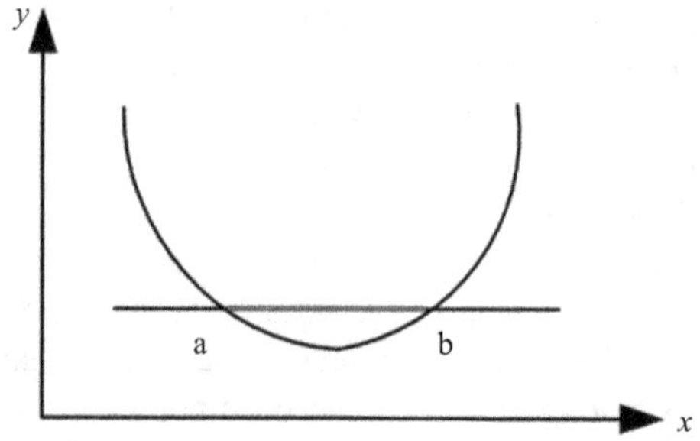

Figure 3.5 Quadratic discriminant function

If choosing an $x \rightarrow y$ mapping, the quadratic discriminant function can be simplified into a linear function concerning y:

$$g(x) = a^T y = \sum_{i=1}^{3} a_i y_i \tag{3.7}$$

and

$$y = [y_1 \ y_2 \ y_3] = [1 \ x \ x^2], \quad a = [a_1 \ a_2 \ a_3] = [c_0 \ c_1 \ c_2] \tag{3.8}$$

where $g(x) = a^T y$ is the generalized linear discriminant function and a is the generalized weight vector.

In general, through appropriate transformations, any higher order discriminant function $g(x)$ can be treated as a generalized linear discriminant function. In this case, $a^T y$ is not a linear function of x, but rather a linear function of y. The equation $a^T y = 0$ defines a hyperplane passing through the origin in the y-space. Similarly, $g(x)$ can be seen as an arbitrary discriminant function expanded as a series and taken as an approximation of its endings. By using this approach, we can leverage the simplicity of linear discriminant functions to solve complex problems. However, this also introduces the challenge of increasing dimensionality.

By designing a linear classifier, we construct a linear discriminant function or a generalized linear discriminant function using the training sample set. The process of designing a linear classifier essentially involves finding better values for w, w_o, or a. The optimal results often occur at the extreme points of the objective function. Thus, the problem of designing a linear classifier transforms into the task of finding the extreme points w^*, w_0^*, or a^* of the objective function using the training sample set.

Common criterion functions are as follows:

- Fisher criterion function
- Perceptual standard function
- Minimum error specifying sample size criterion function
- Least squares error criterion function
- Random minimum error rate linear discriminant criterion function
- Fisher's linear discriminant

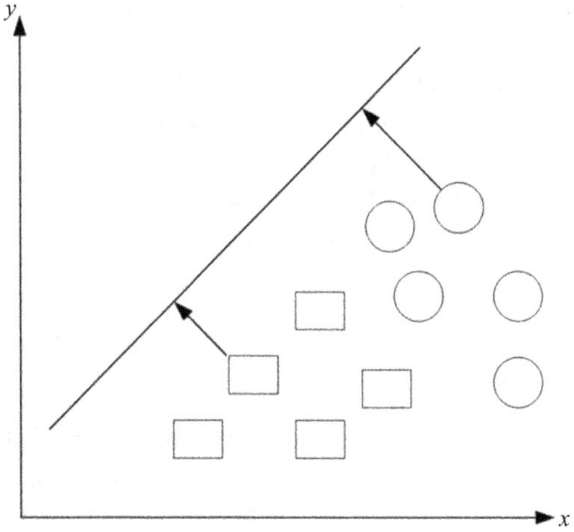

Figure 3.6 Fisher linear discriminant

For example, Fisher's linear discriminant (Figure 3.6) method focuses on the problem of projecting samples from a d-dimensional space onto a straight line to form a 1-D space, that is, compressing the dimensionality to one dimension. However, samples that are well separated in d-dimensional space may be mixed up and indistinguishable when projected into 1-D space. In general, however, it is always possible to find some direction in which the projections of the samples can best be separated into a straight line. The aim is to reduce the dimensionality and partition in a lower dimension.

3.1.2 The optimal classification surface

SVM has evolved from the concept of the optimal classification surface in the case of linear separability, and its basic idea can be illustrated by the 2-D case shown in Figure 3.2. The distance between them is called the classification margin. The optimal classification line is the one that can correctly distinguish between the two classes (a training error rate of 0) and maximize the classification margin. When extended to higher dimensional space, the optimal classification line becomes the optimal classification surface.

Assume the set of linearly divisible samples $(x_i, y_i), i = 1, \ldots, n$, $x \in R^d, y \in \{+1, -1\}$, d-dimensional space under a linearly divisible set of samples: $g(x) = wx + b$, with a classification surface equation of $wx + b = 0$. We can normalize it so that all samples satisfy $|g(x)| \geq 1$, the sample closest to the classification surface satisfies, $|g(x)| = 1$, such that the classification interval is equal to $2/\|w\|$. Thus, the requirement for the maximum classification interval is the requirement for the minimum $\|w\|$. And to require that the classification face correctly classifies all samples is to require that, satisfying the function as below:

$$y_i[wx_i + b] - 1 \geq 0, \; i = 1, \ldots, n, \tag{3.9}$$

Therefore, the classification plane that satisfies the previous formula and minimizes w is the optimal classification plane. The training sample on the hyperplane H_1, H_2 that is parallel to the optimal classification plane and that is the closest point to the classification plane in both classes of samples is the sample that makes the previous equation hold.

The problem of finding the optimal classification surface can be transformed into the following constrained optimization problem:

$$\text{minimize } \frac{1}{2}\|w\|^2 \tag{3.10}$$

$$\text{subject to } y_i\,[wx_i + b] - 1 \geq 0, i = 1, 2, \ldots, l \tag{3.11}$$

It is a quadratic convex programing problem, and since both the objective function and the constraints are convex, there is a unique global minimum solution to this problem according to optimality theory. Applying the Lagrange multiplier method and satisfying the KKT condition:

$$a_i\{y_i[wx_i + b] - 1\} = 0 \tag{3.12}$$

Finally, the optimal classification function to solve the previous problem can be obtained as

$$f(x) = \text{sgn}\{w^* \cdot x + b^*\} = \text{sgn}\left\{\sum\nolimits_{i=1}^{k} a_i^* y_i (x_i \cdot x) + b^*\right\} \tag{3.13}$$

where a^*, b^* are the parameters that determine the optimal division hyperplane, and $(x_i \cdot x)$ is the product of two vector points.

The previous equation shows that the a_i corresponding to the nonsupport vectors are all zero, and the summation is only performed for a small number of support vectors. In the linearly inseparable case, a relaxation term $\epsilon_i \geq 0$ can be added to the condition $y_i[wx_i + b] - 1 \geq 0$ to become as follows:

$$y_i[wx_i + b] + T_i \geq 0 \tag{3.14}$$

Changing the objective to minimizing $\frac{1}{2}\|w\|^2 + C\left(\sum_{i=1}^{n} T_i\right)$, that is, compromising the least misclassified samples and the maximum classification interval, gives the generalized optimal classification surface, where $C \geq 0$ is a constant that controls the degree of penalty for misclassified samples. The optimal classification function obtained in the previous section is as follows:

$$f(x) = \text{sgn}\{w^* \cdot x + b^*\} = \text{sgn}\left\{\sum\nolimits_{i=1}^{k} a_i^* y_i (x_i \cdot x) + b^*\right\} \tag{3.15}$$

This equation contains only the inner product operations of the support vectors in the samples to be classified and the training samples. For a nonlinear problem, a nonlinear transformation can be applied to a linear problem in some higher dimensional space to find the optimal classification surface in the transformation space. Such transformations can be complex, so this idea is not easy to implement in the general case.

According to the theory of general functions, whenever a kernel function $K(x_i, x_j)$ satisfies the mercer condition, it corresponds to an inner product in a certain transformation space. Therefore, a linear classification after a certain nonlinear transformation can be achieved with no increase in computational complexity by using an appropriate inner product function $K(x_i, x_j)$ on the optimal classification surface. The corresponding classification function also becomes

$$f(x) = \operatorname{sgn}\left\{\sum_{i=1}^{k} a_i^* y_i K(x_i \cdot x) + b^*\right\} \tag{3.16}$$

Broadly speaking, an SVM is one that first transforms the input space to a high-dimensional space by a nonlinear transformation defined by an inner product function, in which the optimal classification surface is found. The SVM classification function is formally similar to a neural network, with the output being a linear combination of intermediate nodes, each corresponding to an input sample and an inner product of an SVM, hence the name support vector network.

3.1.3 Selection of kernel functions

Learning machines that implement different types of nonlinear decision surfaces in the input space can be constructed with different kernel functions, leading to different support vector algorithms. In practical problems, the kernel functions are usually given directly. Commonly used kernel functions are as follows:

1. Linear kernel functions:

$$k(x, x_i) = (x \cdot x_i) \tag{3.17}$$

2. Polynomial kernel functions:

$$k(x, x_i) = (s(x \cdot x_i) + c)^d \tag{3.18}$$

3. Radial basis kernel functions:

$$k(x, x_i) = \exp\left(-\gamma |x - x_i|^2\right) \tag{3.19}$$

4. SIGMOD kernel functions:

$$k(x, x_i) = \tanh\left(s(x \cdot x_i) + c\right) \tag{3.20}$$

The nonlinear mapping is the theoretical basis of the SVM method, which replaces the nonlinear mapping to a high-dimensional space with an inner product kernel function. The optimal hyperplane for feature space partitioning is the goal of SVM, and the idea of maximizing the classification margin is the core of the SVM method. The support vector is the training result of the SVM, and the support vector plays a decisive role in the SVM classification decision. SVM is a novel small sample learning method with a solid theoretical foundation. It differs from existing statistical methods in that it does not involve probability measures or the law of large numbers. In essence, it avoids the traditional process of induction to deduction and enables efficient backpropagation from training samples to prediction samples, greatly simplifying the conventional problems of classification and regression. The final decision function of SVM is determined by a small number of support vectors, and the complexity of the computation depends on the number of support vectors rather than the dimensionality of the sample space, which in a sense avoids "dimensional disasters." The small number of support vectors determines the result, which not only helps us to capture key samples and "eliminate" a large number of redundant samples but also predestines the method to be not only algorithmically simple but also "robust."

1. The addition or removal of non-support vector samples does not affect the model.
2. The support vector sample set has some robustness.
3. In some successful applications, SVM methods are insensitive to the choice of kernels.

3.2 MLP/KNN/RF-based approach

Multilayered perceptron (MLP) is a simple-structured neural network. The most common MLP includes three layers: input, hidden, and output, with all three layers of the MLP neural network being completely connected, which is widely used in the detection and diagnosis of the motor drive system. Since its capacity to indirectly detect dynamic nonlinear associations between dependent and independent variables, as well as its ability to detect all possible correlations between predictor variables and multiple training algorithms, the MLP feedforward neural network has been adopted [62]. A three-phase asynchronous motor control method based on radial basis function-MLP (RBF-MLP)-cascaded neural network was proposed in [68]. To obtain rich fault knowledge from stator currents, simple statistical features such as standard deviation, kurtosis, energy, entropy, and variance were extracted. Principal component analysis (PCA) was used to pick the most superior functionality to remove obsolete or irrelevant details and reduce the burden of the classification scheme. The classifier is sufficiently robust, i.e., the classification accuracy does not change in the presence of uniform and Gaussian noise in the input and output. The advantage is that good classification performance can be achieved without the requirement for a large amount of accurate measurement data. Furthermore, MLP was utilized to classify inter-turn short-circuit faults in PMSM stator windings at different speeds, loads, and fault severity. These states led to the complexity of the fault features. Depending on the complexity, the nonlinear relationships between the relevant features are identified to detect different severity levels of inter-turn short circuits.

Due to its easy implementation and substantial classification efficiency [69], the *k*-nearest neighbors (KNN) method is a popular classification method in data mining and statistics. The principle structure of KNN is as shown in Figure 3.7. Reference [70] presented a system for detecting bearing faults in electric motors and monitoring the bearing loss. The approach used spectral kurtosis (SK) and

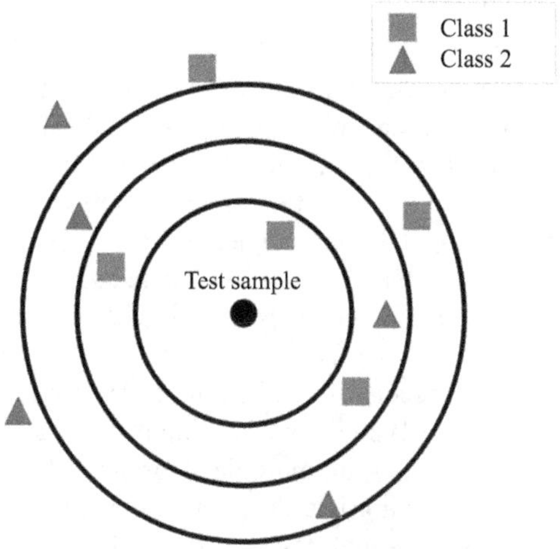

Figure 3.7 The structure of KNN

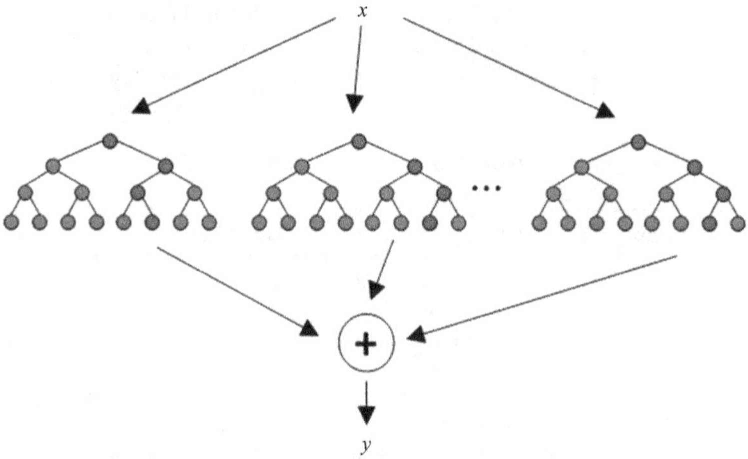

Figure 3.8 The basic structure of the random forest

reciprocal association to extract fault features, which were then combined PCA and semi-supervised KNN distances to provide health metrics.

Random forest (RF) models are based on the classification and regression of tree groups (CART) [71]. RF is a method that uses integrated learning to combine numerous trees, with the fundamental unit becoming a decision tree, each of which is a classifier. The categorization results for an input sample will then be *N* trees. The RF integrates all classification votes and designates the most voted category as the final output, which is a basic bagging idea. Figure 3.8 illustrates the basic structure of the RF. An RF classifier was proposed in [72] for the classification of bearing faults. The statistical features of the bearing vibration signals were computed and fed to the RF and ANN classifiers. Reference [73] developed a hybrid model that combines a fuzzy min–max (FMM) neural network with CART for online motion detection and diagnosis tasks. In addition to strong online performance, the FMM-CART produced helpful decision trees to explain the collected functional information.

To sum up, the ML methods mentioned earlier are all shallow models with simple structures, high computational efficiency, and great classification performance. Table 3.1 summarizes the classical ML methods for fault classification.

MLP is a feedforward neural network consisting of multiple layers of neurons. It is one of the basic models of DL and is widely used in various ML tasks such as image classification, speech recognition, and natural language processing.

The basic structure of MLP consists of an input layer, a hidden layer, and an output layer. Each layer consists of multiple neurons, which are connected by connection weights. The following is a detailed explanation of MLP. Input layer: This layer receives the raw data and passes it on to the hidden layer. Each input neuron corresponds to a feature of the data, such as a pixel value in an image classification task. Hidden layer: This layer is the core component of the MLP and is responsible for learning and extracting high-level feature representations from the input data; the MLP can have one or more hidden layers, each containing multiple neurons. The neurons in the hidden layers are not connected, nor are the neurons in the same layer connected. Output layer: This layer receives the

Table 3.1 Highlight of classical machine learning method

Reference	Theme	Principle	Highlight
[63]	MEISVM	• Sensor information correlation • Pairwise secondary optimization	• Multi-fault diagnosis • Highly rely on input feature data
[64]	mRMR GSSVM	• Maximize feature relevance • Minimize redundancy between • Features	• Fast calculation and high robustness • Random parameter setting
[65]	CMFE ESVM	Obtain vector similarity	Sensitive to nonlinear feature
[67]	FFT SVM	Structural risk minimization	Acceleration signal analysis application
[68]	RBF MLP	• Competition rules and metric • Combinations	• Low demand for accurate measurement data • High classification performance
[70]	SK KNN	Category distance calculation	To improve speed and space efficiency
[71–73]	RF	Random sampling and replacement	• Effective classification ability • Insufficient generalization ability

representation from the hidden layer and generates the final output. The number of neurons in the output layer depends on the nature of the task, for example, a classification task may have as many output neurons as there are categories, with each neuron representing the probability of a category. Weights: Each connection has an associated weight, which indicates the strength of the connection between neurons. The weights determine how the input signal is propagated and combined between the neurons and are adjusted by the learning process. Activation function: The activation function is a nonlinear transformation function for each neuron in the MLP. It allows the MLP to learn and represent more complex functional relationships by introducing nonlinear elements. Common activation functions include sigmoid, ReLU, tanh, and others.

Feedforward: The MLP is a feedforward neural network. The process of passing input data from the input layer through the hidden layer to the output layer is called feedforward. In the feedforward process, each neuron calculates a weighted sum of its input signals and produces output through an activation function. Backpropagation: It is the key step in training an MLP. It uses the difference between the labels of the training data and the model output to calculate the loss and minimizes it by adjusting the weights. The backpropagation algorithm uses a gradient descent approach to update the weights to progressively optimize the performance of the MLP.

3.2.1 Model of a single-layered perceptron

As shown in Figure 3.9, single-layered perceptrons are those with only one layer of processing units. The input layer is the perceptual layer, and these n nodes are only responsible for bringing in external information. The output layer is the processing layer, with m neuron nodes, all of which have processing capabilities. The m nodes output processing information to the outside, constituting the output column vector. The two layers are connected by a vector of weights, the positive and negative of which reflect the excitation (excitable pulses) and inhibition of the neurons.

$$\begin{aligned} X &= (x_1, x_2, \ldots, x_i, \ldots, x_n)^T \\ O &= (o_1, o_2, \ldots, o_i, \ldots, o_n)^T \\ W_j &= (w_{1j}, w_{2j}, \ldots, w_{ij}, \ldots, w_{nj})^T \quad j = 1, 2, \ldots, m \end{aligned} \tag{3.21}$$

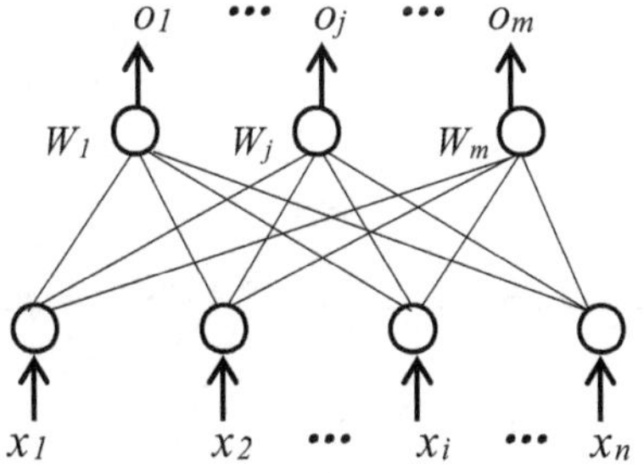

Figure 3.9 Single-layered perceptron

Net input:

$$net_j = \sum_{i=1}^{n} w_{ij}x_i \tag{3.22}$$

Weighted summation of the *n* inputs to neuron *j*.
Output:

$$o_j = \text{sgn}(net_j - T_j) = \text{sgn}\left(\sum_{i=0}^{n} w_{ij}x_i\right) = \text{sgn}\left(W_j^T X\right) \tag{3.23}$$

$$-T_j = w_{0j}x_0$$

$$w_0 = T_j$$

$$x_0 = -1 \tag{3.24}$$

The simplest single computational node perceptron has a classification function (Figures 3.10–3.13). The classification principle is that classification knowledge is stored in the perceptron's weight vector (which contains thresholds), and the classification judgment interface determined by the weight vector classifies the input patterns into two categories:

Logical "with" functions with perceptrons
Logical "with" truth tables

x_1	x_2	Y
0	0	0
0	1	0
1	0	0
1	1	1

$$w_1x_1 + w_2x_2 - T = 0$$

$$0.5x_1 + 0.5x_2 - 0.75 = 0 \tag{3.25}$$

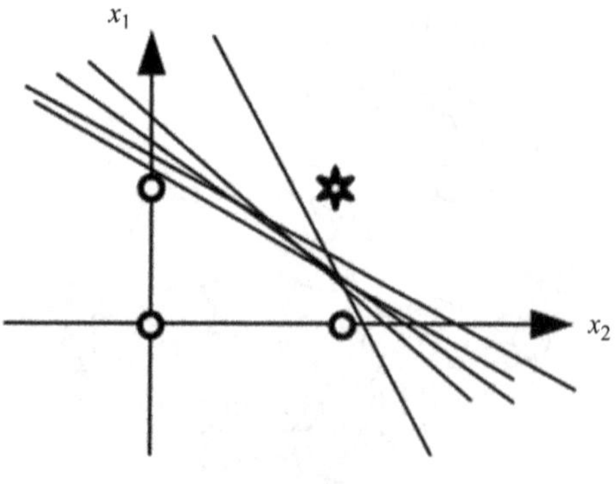

Figure 3.10 Classification performance of logical "with" functions with perceptrons

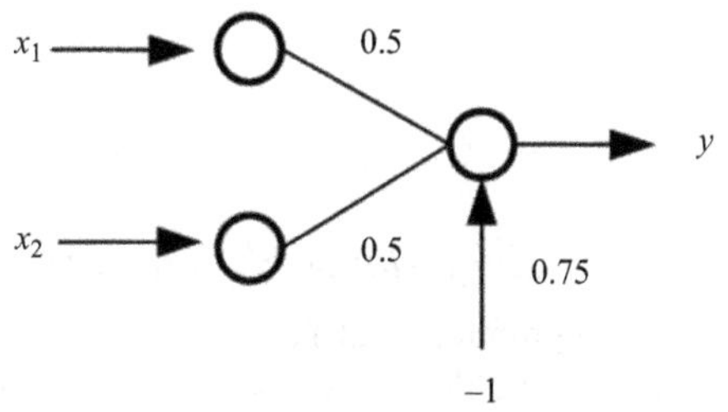

Figure 3.11 Perceptron structure

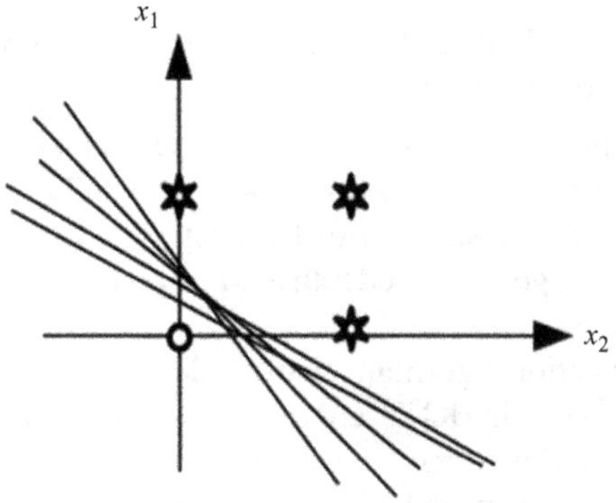

Figure 3.12 Classification performance of logical "or" functions with perceptrons

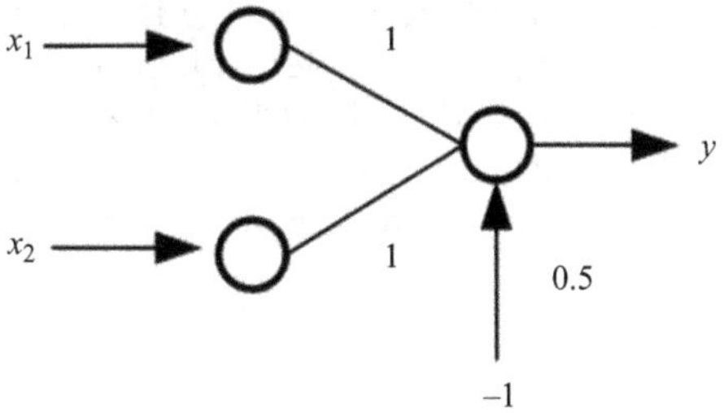

Figure 3.13 Diagram of connection weight

Implementing logical "or" functions with perceptrons
Logical "or" truth tables

x_1	x_2	y
0	0	0
0	1	1
1	0	1
1	1	1

$$w_1x_1 + w_2x_2 - T = 0$$
$$x_1 + x_2 - 0.5 = 0 \tag{3.26}$$

3.2.2 Limitations of the single-layered perceptron

The limitation of the single computational layer perceptron is that it only has the ability to classify linearly separable problems (Figure 3.14). In reality, a large number of linear problems are linearly indistinguishable. The "is-or" problem is such a problem. The key to solving this problem is to introduce a hidden layer as an internal representation of the input pattern.

3.2.3 Nearest neighbor (k-nearest neighbor (KNN)) classification algorithm

Classification is an important technique in the field of data mining that finds a classification model from a set of known training samples and uses this classification model to predict the samples to be classified. It is essential to build an effective classification algorithm model that will eventually process the samples to be classified.

The main classification algorithms in use today are naïve Bayes, SVMs, KNN, nets, and decision trees. The KNN algorithm is a theoretically mature method, originally proposed by Cover and Hart in 1968, with a very simple and intuitive idea that is easy to implement quickly. As a result, the KNN algorithm has been widely used in fields such as automatic Chinese text classification for its simplicity of implementation and high classification accuracy. The distance between the sample to be classified X and each training sample is calculated according to the distance function (as similarity), the k samples with the smallest distance from the sample to be classified are selected as the KNNs of X, and finally, the class to which the majority of the KNNs of X belong is taken as the class of X. KNN is arguably one of the most straightforward methods used to classify unknown data. In

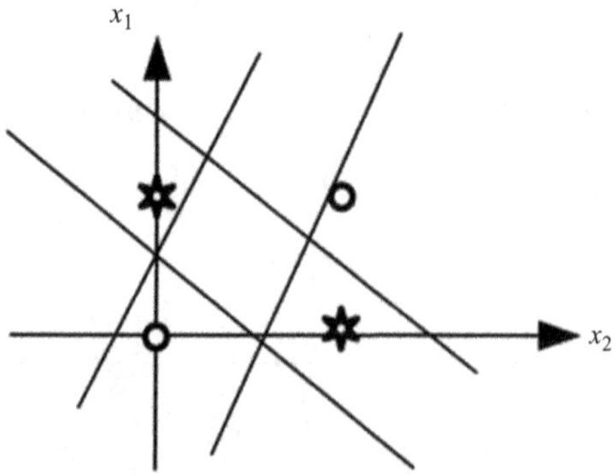

Figure 3.14 Classification performance of logical "xor" functions with perceptrons

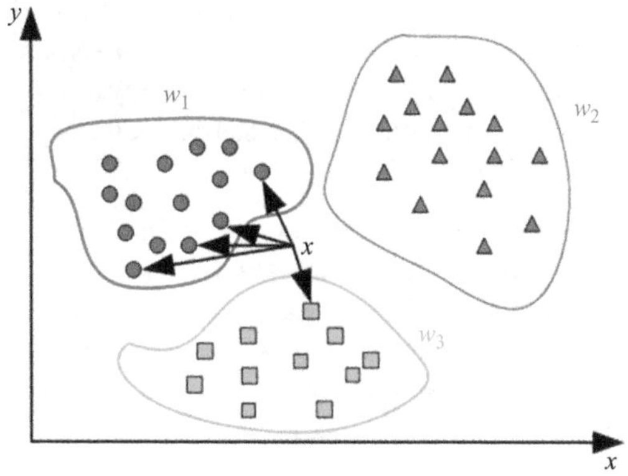

Figure 3.15 Classification performance

simple terms, a KNN can be thought of as having that pile of data that you already know the classification of, and then when new data comes in, start finding the distance to each point in the training data and then pick the k points closest to that data to see what type those k points belong to and then use the principle of majority rule to classify the new data. Figure 3.15 indicates the basic classification performance pattern of KNN.

3.2.3.1 Advantages and disadvantages of KNN

Advantages:

1. The algorithm is relatively simple in idea and easy to implement.
2. No need to retrain when new samples are to be added to the training set (i.e., low cost of retraining).
3. Computational time and space are linear in the size of the training set (which is not too large in some cases).

Shortcomings:

1. Slow classification speed:
   The time complexity and storage space of the KNN algorithm increases rapidly with the size of the training set and the number of feature dimensions. This is because each time a new sample is classified has to be computed together with all the training sets to compare the similarity to take out the top k-classified samples. The time complexity of the whole algorithm can be expressed as O ($m \times n$), where m is the number of feature items (attributes) selected, and n is the number of samples in the training set.
2. Equal weights for each attribute, which affects accuracy:
   When samples are unbalanced, such as when one class has a large sample size and the other classes have a small sample size, it is possible that when a new sample is entered, the sample from the large-capacity class will be the majority of the k neighbors of that sample. The algorithm only counts the "nearest"

neighbor samples, so if a class has a large number of samples, then the target sample may not be close to that class, but the target sample will be classified under that class, affecting the classification accuracy.
3. Strong dependence on the volume of the sample pool.
4. Poor determination of the *k*-value:

If the k value is chosen too small, the number of nearest neighbors obtained is too small, which will reduce the classification accuracy and also amplify the interference of noisy data, while if the *k* value is chosen too large if the sample is to be classified belongs to a class containing fewer data in the training set, then when the KNNs are chosen, data that are not similar are also included, causing an increase in noise and resulting in a reduction in the classification effect.

3.2.4 Some improvement strategies for KNN

1. From the perspective of reducing the computational complexity:
   When the sample size is large as well as the number of feature attributes, the efficiency of KNN algorithm classification will be greatly reduced. The following method can be used for improvement. If the attributes of the samples are reduced before using the KNN algorithm, and those attributes that have less influence (less important) on the classification result are removed, the KNN algorithm can be used to quickly derive the class of the samples to be classified, and thus better results can be obtained. Rough set theory, when used for attribute reduction in decision tables, can remove irrelevant and redundant attributes from a decision table while keeping the decision capabilities in the table unchanged. The method of narrowing down the training sample: deleting a part of the original sample that is not very relevant to the classification and using the remaining sample as a new training sample or selecting some representative samples from the original training sample set as a new training sample.

   By clustering (clustering), the centroids resulting from the clustering are used as new training samples.
2. From the perspective of optimizing the similarity measure:
   The basic KNN algorithm is based on the Euclidean distance to calculate the similarity of samples, and this method is very sensitive to noisy features.

   To change the shortcomings of the traditional KNN algorithm where the features play the same role, different weights can be given to the features in the distance formula for measuring similarity, and the weights of the features are generally set according to the role of each feature in the classification.
3. From the perspective of optimizing the decision strategy:
   The disadvantage of the decision rule of the traditional KNN algorithm is that when the distribution of samples is uneven (the number of training samples is not balanced between categories, or even if the basic number is close, due to the difference in the size of the area they occupy), following only the first *k* neighboring orders without considering their distances can cause misclassification and affect the performance of classification. This can be improved by homogenizing the sample distribution density.

4. From the perspective of selecting the appropriate k-value:
   Since almost all the calculations in the KNN algorithm occur in the classification phase, the classification effect relies heavily on the selection of the k-value. And so far, a better way of selecting k-values can only be adjusted by repeated trials.

3.2.5 Random forest

RF is an ensemble learning method that combines multiple decision trees to perform classification and regression tasks. It is a powerful ML algorithm proposed by Leo Breiman and Adele Cutler. A decision tree is a structure composed of nodes and edges. Each node represents a feature or attribute, and edges represent possible values for that feature. Decision trees use a sequence of judgments and decisions to classify or regress input data to make predictions. RF is an ensemble learning method that builds multiple decision trees and ensembles them to improve the accuracy and robustness of the model. By combining predictions from multiple models, RFs can reduce the risk of overfitting a single decision tree and provide better generalization. RFs introduce two sources of randomness, random sampling, and random feature selection. With random sampling, each decision tree is sampled back and forth from the training data, which means that each tree has the potential to get a different training set. Random feature selection means that when building a decision tree, only some features are considered in each division of a node, rather than all features, thereby increasing the variability between trees. RFs use Gini impurity to evaluate the ability of feature division while building decision trees. Gini impurity measures the probability that a randomly selected sample will be misclassified during classification. At each node, RF selects the feature with the lowest Gini impurity for classification. For classification tasks, each decision tree in the RF predicts the input samples and gives a class label. The final prediction is obtained by voting or averaging the predictions of all decision trees. For regression tasks, the RF's prediction is the average of all decision tree predictions. RFs are very robust and can handle many input features and samples. RF is an algorithm that integrates multiple trees through the idea of ensemble learning. Its basic unit is a decision tree, which is essentially a branch of ML, that is, ensemble learning. There are two keywords in the name of RF, one is "random" and the other is "forest." One tree can be called a forest, hundreds of trees can be called a forest, and that is an apt metaphor.

1. The concepts of information, entropy, and information gain:

These three basic concepts are fundamental to decision trees and are the basis for determining the order in which features are selected when they are used to classify a decision tree. Once you understand them, you will have a general understanding of decision trees.

To quote Shannon, information is something that is used to eliminate random uncertainty. Of course, this is a classic saying, but it is still hard to figure out what this stuff looks like, and it may mean different things in different places. For a decision tree in ML, if the set of things with categories can be divided into multiple classes, the information of a class x_i can be defined as follows.

$I(X = x_i) = -\log_2 p(x_i)$ is used to denote information about a random variable, and $p(x_i)$ refers to the probability of when x_i occurs. Entropy is used as a measure of uncertainty; the greater the entropy, the greater the uncertainty in $X = x_i$ and vice versa. For a classification problem in ML, the higher the entropy, the greater the uncertainty in the category, and vice versa. Information gain is used as a metric for feature selection in decision tree algorithms; the greater the information gain, the better the selectivity of the feature.

The decision tree algorithm is a method for approximating the value of a discrete function. It is a typical classification method that first processes the data, generates readable rules and a decision tree using an induction algorithm, and then uses the decisions to analyze the new data. Essentially, a decision tree is a process of classifying data through a set of rules.

3.2.6 *Awareness of the CART algorithm*

The CART is an implementation of the decision tree. The CART algorithm is a binary-recursive partitioning technique that divides the current sample into two subsamples so that each non-leaf node generated has two branches; thus the decision tree generated by the CART algorithm is a simple binary tree. As the CART algorithm is a binary tree, it can only take a "yes" or "no" decision at each step, even if a feature has more than one value, and the data are divided into two parts. In the CART algorithm, there are two main steps:

1. Recursive division of the sample into tree-building processes
2. Pruning with validation data

   The CART algorithm divides the dimensional space into nonoverlapping rectangles in a recursive manner. The division steps are roughly as follows: (1) Select an independent variable, and then a value chosen to divide the dimensional space into two parts, one with all points satisfied and the other with all points satisfied, for noncontinuous variables the attribute values take only two values, that is, equal to the value or not equal to the value. (2) Recursively process the two parts obtained above by re-picking an attribute as in step (1) and continuing to divide the dimensional space until the whole dimensional space has been divided.

CART principle: Using the Gini coefficient as an indicator to build a decision tree, the probability distributions of the various values of the labeled columns after various level splits are calculated separately:

$$Gini(A) = 1 - \sum_{i=1}^{C} p_i^2 \tag{3.27}$$

where p_i denotes the probability of belonging to class i. When $Gini(A) = 0$, all samples belong to the same class.

3.2.7 *Random forest generation*

There are many classification trees in an RF. To classify a classification tree, samples need to be fed into each tree to be classified.

Bootstrap sampling method: with put-back sampling.

The idea of bagging in an RF: The classification results of several weak classifiers (multiple trees) are voted on to form a strong classifier.

3.2.8 Random forest generation rules

1. If the size of the training set is N, N training samples are randomly and with put-back drawn from the training set to form a new set of samples (with duplicate training samples).
2. If the feature dimension of each sample is, specify a constant m, and $m < M$, randomly select a subset of m features from the M features, and each time the tree splits, select the best one from these m features, with the size of m remaining constant during the growth of the decision tree. (Note: No put-back selection is used here.)
3. Each tree is grown as maximally (completely) as possible, and without pruning processes.

The previous three steps give a decision tree. Repeat this process X times to get X decision trees. A test sample is then classified once, giving X classification results. Use a simple voting mechanism (or use the final classification result) to determine which class the sample belongs to.
Note: The introduction of two classes of randomness is crucial to the generation of the RF.
Two factors determine the effectiveness (error rate) of random forest classification

1. Correlation of any two trees in the forest: The greater the correlation, the greater the error rate.
2. Classification ability of each tree in the forest: The stronger the classification ability of each tree, the lower the error rate of the whole forest.

Decreasing m (the number of features selected) decreases the relevance and classification ability of the trees; increasing m increases both. So the key question is how to choose the optimal m (or range), which is the only parameter of the RF.

3.2.9 Out-of-bag error rate (OOB error)

As we mentioned earlier, the key problem in constructing an RF is how to choose the optimal m. The solution to this problem is mainly based on calculating the out-of-bag (OOB) error rate. In constructing each tree, we use random and put-back sampling for the training set. Therefore, for each tree, about 1/3 of the training instances are not involved in the tree generation, which is called OOB sample data of the tree. These data are not involved in the fitting of the training model and can therefore be used to check the generalization ability of the model. And such a sampling characteristic allows for OOB estimation, which is computed as follows:

1. For each sample, calculate how well it is classified by the tree it is an OOB sample from (about $1/3$ of the tree).
2. Then a simple majority vote is used as the classification result for that sample.

3. Finally, the ratio of the number of misclassifications to the total number of samples is used as the OOB error rate for the RF.
4. The OOB error rate is an unbiased estimate of the RF generalization error, and its results approximate a k-fold cross-validation that requires extensive computation. So there is no need to cross-validate it or to use an independent test set to obtain an unbiased estimate of the error. It can be evaluated internally, that is, an unbiased estimate of the error can be created during the generation process. Once we know how the OOB is calculated, we can know the optimal m by picking different ms, calculating the OOB error, and finding the value of m that corresponds to the smallest OOB error.

Characteristics and advantages of RFs include the following:

1. The introduction of two forms of randomness reduces the risk of overfitting, making RFs less prone to overfitting.
2. The introduction of two forms of randomness also enhances the noise robustness of RFs, making them capable of handling noisy data effectively.
3. RFs are not sensitive to default parameter values due to the two forms of randomness employed.
4. RFs have good adaptability to the dataset. They can handle both discrete and continuous data, and there is no need for data normalization. They can efficiently operate on large datasets.
5. RFs can handle input samples with high-dimensional features without requiring dimensionality reduction techniques.

3.2.10 Overfitting

Overcomplicating a hypothesis to obtain a consistent hypothesis is called overfitting. In ML, assumptions are overcomplicated to obtain an accurate result. Figure 3.16 indicates the overfitting in different error.

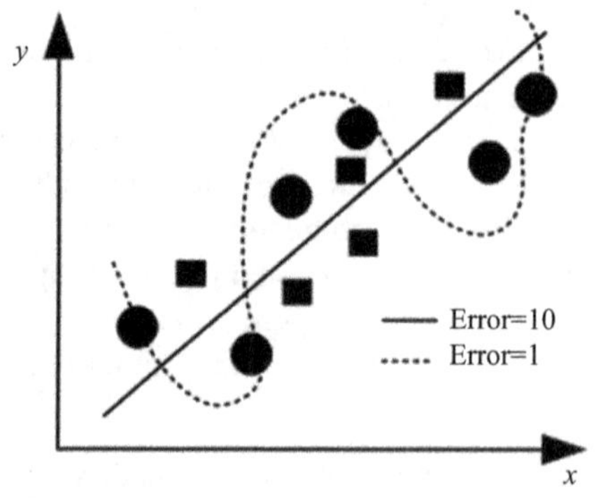

Figure 3.16 Overfitting

3.2.11 Why is a random forest effective in avoiding overfitting?

An RF is made up of many trees grouped. Each tree alone is overfitted, but, since it is overfitted, it will fit small details, so an RF tries to avoid overfitting by introducing randomness so that each tree fits different details, and by grouping the trees, the overfitting part is eliminated.

3.2.11.1 Disadvantages

1. Overfitting in some noisy classification or regression problems.
2. For data with different levels of attributes, attributes with more levels of classification will have a greater impact on the RF, so the weights of the attributes produced by the RF on such data are not reliable.

In contrast to traditional ML methods with manual feature labeling, a large amount of research has focused on how to extract representative features from the original signal. On the other hand, the extracted data may contain redundant or insensitive data. To identify sensitive characteristics, certain dimensionality reduction methods were used, which may have an impact on diagnostic findings as well as computing performance. Most of the studies on intelligent fault diagnosis have produced valuable results, but there are still two obvious drawbacks. (1) Manual extraction of features requires a priori knowledge, which requires a lot of practical work to determine and may have greater checking popularity. (2) Traditional ML techniques cannot effectively distinguish complex information in raw data, and the shallow structure of artificial neural networks limits their ability to understand the complex nonlinear relationships hidden in the measurement data. In terms of a large number of hidden neurons, DNNs can obtain nonlinear representations of data. It has achieved higher performance in the field of motor FDD.

3.3 Convolutional neural network

Convolutional neural network (CNN) is a feedforward neural network that mainly stimulates the activity of the visual system of the human brain. Local receptive fields, weight sharing, and spatial domain secondary sampling are the three main architectural principles in the structure of CNN. Therefore, CNN is well suited to processing 2-D data like images. CL, pooling layers (PLs), completely connected layers (FL), and SoftMax layers are the four kinds of layers that make up CNN. The following equation can be used to mathematically model the operation of processing input data in the CL:

$$x_j^i = f\left(\sum_{i \in M_j} x_i^{l-1} * k_j^l + b_j^l\right) \tag{3.28}$$

where x_i^{l-1} indicates the input data of layer l, * denotes the convolution operation. The layer is made up of n kernels, each with its weight matrix and bias vector.

Because of its versatility, the rectified linear unit (ReLU) is often used as a CL activation function. The equation for ReLU is as follows:

$$f(y) = (0, y) \tag{3.29}$$

The following equation can be used to express PL mathematical operation:

$$x_j^{l+1} = s\left(x_j^l\right) \tag{3.30}$$

After the nonlinear combination of features extracted by FL, the SoftMax layer acts as a classifier. The SoftMax function converts an N-dimensional real vector into a set of real vectors (0, 1). The SoftMax function has the following equation:

$$p_i = \frac{\mathrm{expexp}(z_i)}{\sum_{j=1}^{N} \mathrm{expexp}(z_j)} \tag{3.31}$$

Signal-processing techniques with different characteristics are combined with CNN for fault signal diagnosis [74,75]. Reference [74] proposed a novel motion state monitoring framework with an adaptive implementation of 1-D CNN, where the two main modules of traditional fault detection steps, feature extraction and classification, were fused into one. The proposed approach extracts the best features through proper training. Therefore, manual parameter adjustment and manual feature extraction are not required, achieving more efficient fault diagnosis capability. Reference [76] proposed an integrated strategy based on data-driven and DL to deal with initial faults. The average moving technique was introduced into the typical correlation analysis (CCA) framework, making the new residual signal more sensitive to initial faults. Moreover, new test statistics that work closely with Kullback–Leibler divergence (KLD) were proposed from a probabilistic perspective and greatly improved fault detection performance. The fault matrix was defined and used as the input of the CNN, whose feature extraction capability was greatly improved compared with the conventional method, which helped to diagnose the initial faults accurately. The CNN shown in Figure 3.17 was used as a construction block multi-signal framework. Randomly initializing the designed

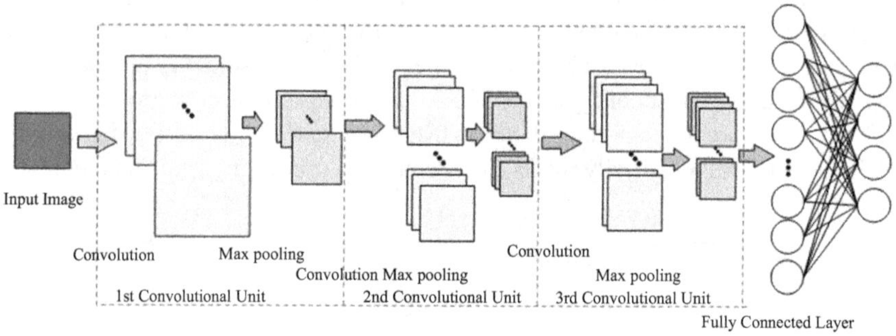

Figure 3.17 The basic structure diagram of CNN

model, the deep model was trained using the training dataset, minimizing the error between model output predictions and actual labels by iteratively updating the parameters of the DCNN model [77]. The proposed method has one input and output layer, three hidden convolutional units, each of which was followed by a max-pooling operation, and one fully connected hidden layer following the convolutional and max-PLs. ReLu was implemented as an activation function, continuous wavelet transform (CWT) was used to convert the time-domain signal into grayscale images as input. The model was trained and verified with experimental data, including current and vibration signals. The training time was 201s and 156s, and the accuracy rates of 98.72% and 98.26% were obtained, respectively. In addition, CNN combined with other networks has been proposed. A convolutional recurrent neural network (RNN) was used to diagnose multiple faults of high-speed train bogies by combining CNN and RNN [78]. The model inherited the functions of both CNN and RNN. A new approach was proposed that combines CNN and extreme learning machine (ELM). CNN demonstrated strong automated feature extraction capabilities, while ELM was proposed as a quick and efficient classification algorithm [79]. To improve the feature learning capabilities, a CNN with a square pool configuration was built and used as an automated feature extractor in the first level. ELM was further used in the second stage to increase classification accuracy and learning speed. In [80,81], the initial vibration signal was input to a deep CNN, named deep CNNs, with wide first-layer kernels (WDCNN). A large first-layered convolutional kernel and a deep network structure with narrow CLs were the two key features of the WDCNN. The proposed model enhances the accuracy of the current CNN fault diagnosis. Besides, a new model named convolution neural networks with training interference (TICNN) was proposed to solve the fault diagnosis problem in [82]. Without the requirements for a time-consuming manual feature extraction process, TICNN processed the raw vibration signal directly. Without any time-consuming denoising preprocessing, the raw time signal was used as data. Meanwhile, the model was independent of any domain adaptation algorithm or target domain information. An online fault diagnosis scheme was proposed in [83]. Conventional ANNs for fault diagnosis classification tasks were trained offline with historical data, and then the trained model was used for online fault diagnosis. Due to the limitations imposed by the time consumption associated with the training process of the model, an online fault diagnosis algorithm, including two phases was proposed. In the first stage, an SVM was used to separate the healthy data from the faulty data, and in the second stage, a CNN was trained to learn the features used to isolate the faults. Although the proposed method was an end-to-end self-supervised learning model, the performance of the CNN in the second stage was limited by the number of datasets and the training time consumption. The data were used only in the healthy state through the fault diagnosis function in self-supervised learning. The operating condition or operating condition class of the powertrain was defined in the first stage by using a class of SVMs. The generated health classes were used to train a CNN-based classifier. This approach outperformed current algorithms and approaches that use domain feature extraction. A new LeNet-5-based CNN for fault diagnosis was proposed in [84], three datasets were used to verify

CNN models based on LeNet-5 with 8 layers, which comprised three transformed pictures of 64×64, 64×64, and 16×16. On three separate datasets, the presented CNN had prediction accuracies of 99.79%, 99.48%, and 100%. By converting the signal into a 2-D image, the method extracted the features of the converted 2-D image and eliminates the influence of artificial features. The novel data preprocessing approach presented in the literature transformed raw time-domain signal data into 2-D grayscale pictures without the use of any predetermined parameters, removing as much as possible the knowledge of the expert. However, the limitations of the method in terms of practical applicability included the following aspects. First, the most prevalent fault circumstances must be identified, which may be achieved using a dictionary list type. Otherwise, unknown faults will be categorized incorrectly as known faults. Second, the training procedure took a long time. TL can be used to minimize training time in the future based on these constraints.

Figure 3.18 indicates the development of CNN based algorithm. CNNs are DL structures inspired by visual perception mechanisms, discovered in 1959 by Hubel and Wiesel in animals. In 1959, Hubel and Wiesel discovered that the visual cortex cells of animals were responsible for detecting light in the perceptual field. Inspired by this, in 1980 Kunihiko Fukushima proposed a new cognitive machine that is considered to be the forerunner of CNNs. In 1990, LeCun published an important article that laid down the structure of the present CNN. They constructed a multi-layered feedforward neural network called LeNet-5 and used it for handwriting recognition. Like other feedforward neural networks, it can be trained using a backpropagation algorithm to train it. It is effective because it learns effective features from the original image with little or no preprocessing of the image. It is effective because it learns valid features from the original image with little or no preprocessing of the image. However, LeNet-5 was unable to perform well on complex problems such as large-scale images due to the lack of training data and the computing power available at the time. However, LeNet-5 was unable to perform well on complex problems such as large-scale image classification and video classification processing due to the lack of training data and insufficient computing power at the time. Since 2006, many more methods have been proposed to overcome the difficulties encountered in training difficulties encountered when training deep CNNs. One of the important ones is the AlexNet network, which is very similar to LeNet-5, but with very significant improvements in image classification tasks. It is very similar to LeNet-5 but with very significant improvements for the image classification task. With the success of AlexNet, many new neural networks

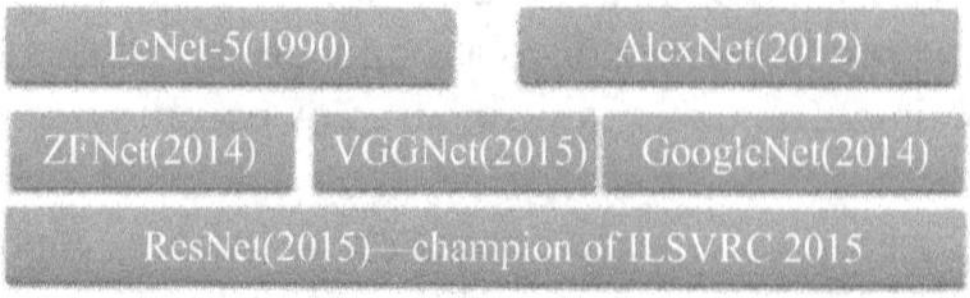

Figure 3.18 Development of CNN-based algorithm

have been proposed to improve it. On the basis of this, many new neural networks have been proposed to improve its performance.

A very important trend in these neural networks is DEEPER, with ResNet being 20 times deeper than AlexNet and VGGNet being 8 times deeper. As the depth increases, the network can better approach the nonlinear target equation and can learn more effective features. Of course, this also increases the complexity of the model, making it difficult to train and prone to overfitting. As a result, several approaches have been proposed to solve the training and overfitting problems of CNNs.

Similar to traditional feedforward neural networks, several factors determine the model of the entire network. While the hidden layers of our traditional feedforward neural networks are generally fully connected, the hidden layers of CNNs are further divided into CLs, PLs, and a final hidden layer. The distinction between these hidden layers is an important part of what makes a CNN. The choice of activation function for each neural element and the choice of loss function for the whole network are similar to those of FNN. CNNs also have a wide range of applications, including image classification, target detection, target recognition, target tracking, text detection and recognition, and location estimation, to name a few. The important difference between CNNs and FNNs mentioned earlier is the difference in the hidden layer. A hidden layer is one in which there is a fully connected layer, which then forms a CNN together with the final fully connected layer.

An important operation of CNN, convolution, is the core idea of CNN. It is convolution that effectively extracts image features for use in post image recognition. Figure 3.19 is a schematic representation of a two-dimensional convolution, and since it is a discrete convolution, the convolution can be directly understood as a matrix. Since it is a discrete convolution, the convolution can be directly understood as a matrix multiplication, that is, the multiplication of two matrices, one being the input matrix and the other being the output matrix. One is the input matrix and the other is the convolution kernel matrix. The input matrix generally represents a 2-D. The input matrix generally represents a 2-D input image, while the convolution kernel can be understood as an operator in image processing. In image processing, for example, these operators can implement some edge detection or Gaussian blur. Edge detection or Gaussian blur, then the convolution operation can be understood as some kind of feature processing of an image. As shown in the Figure 3.19, a convolution operation refers to a convolution kernel with an image matrix of the same image matrix of the same size, and then sliding the convolution kernel down or the kernel is then slid down or to the right for the next convolution operation. This is done by sliding the kernel over the entire image. This is done by sliding the kernel over the entire image to create a convolution layer.

The size of the convolution kernel on the right is 2 by 2. Roberts' edge detection operation with a 2×2 matrix doing convolution in the input on the left is equivalent to edge detection at point e. The 2×2 matrix doing the convolution operation is equivalent to edge detection at point e. The Roberts operator is one of the simplest operators and it uses the local difference operator to find edge information, which uses the diagonal direction of two neighboring pixels'

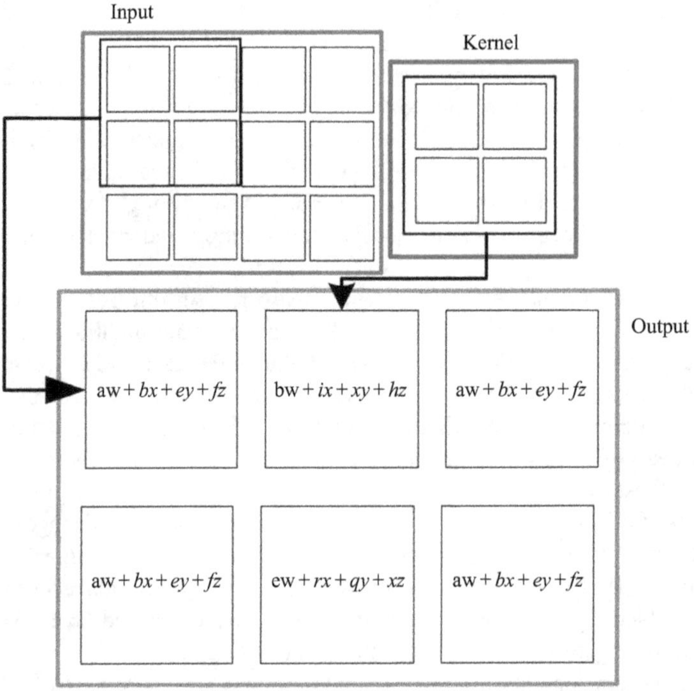

Figure 3.19 Convolution operation

difference to approximate the gradient of the detected edge. The Roberts operator here is only a first-order operator and still extracts very limited edge information, there are other second-order operators such as, for example, the Laplace operator. In addition, the Roberts operator here only extracts the diagonal gradients, not the vertical and horizontal gradients. And horizontal directions, so other operators can be used to extract gradients in multiple directions, for example, other operators can be used to extract gradients in multiple directions, such as the Sobel operator, the Prewitt operator, and so on.

3.3.1 Sparse connectivity

It is the presence of convolutional kernels and the fact that the size of a convolutional kernel is smaller than the whole image that is why the idea of sparse concatenation has emerged. In Figure 3.20, our traditional FNNs are generally fully connected, as shown in the blue box at the bottom right. The blue box on the right shows that although this is 1-D, it is the same as 2-D. We can think of a convolution kernel as a connected right where the convolution kernel is the same size as the image, which is equivalent to a full connection, so the w in the output layer and the s_3 in the output layer receive all the neurons in the input layer. Looking again at the connections in the red box, we can imagine the image as 2-D, where only 1-D

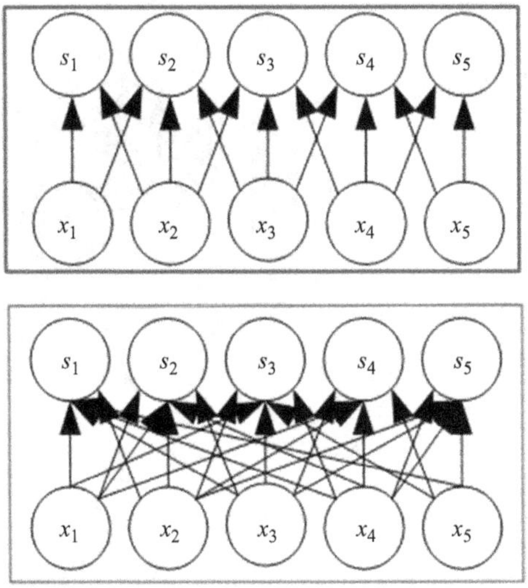

Figure 3.20 Sparse connectivity

information is shown, and the size of the convolution kernel is 3 times the size of 3, so this creates a sparsity of connection weights, that is, ultimately s_3 is only affected by three neurons. This makes sense, that is, according to some biological studies. Our visual perceptual cells are locally perceptive, for example, we use a very small convolutional kernel to extract information about the layout edges of the whole image. This is when it makes little sense to use full connectivity to learn and the complexity of the model is very high.

The first part of the descent to the convolution kernel sliding over the whole image is weight sharing, as the convolution kernel here represents the weight, and this weight is used for the whole image. This means that for each local position of the image, they are connected to the next layer of the network. For each local position of the image, they are connected to the next layer with the same weight. This is why there is also a name for weight sharing, *tied* weights.

The black arrow in the red box as in Figure 3.21 indicates an element of a 3-D convolutional kernel in a convolutional network. Since the weights in the left figure are shared, resulting in that for all input neural elements, the value of the black arrow is the same. In contrast, in the blue box, the black arrow indicates the most central weight in the fully connected layer. The central weight, which was used only once, was not shared again by any of the neurons was shared again. Now that we understand the concepts of convolutional operations and sparse connections and weight sharing, it is possible to generate CLs from these generated from these. Due to weight sharing, that is, one convolutional kernel can only extract features from one image, to extract multiple features from one image, multiple convolutional kernels are required. To extract multiple features of an image, multiple convolutional kernels are required, each of which generates a feature map, as

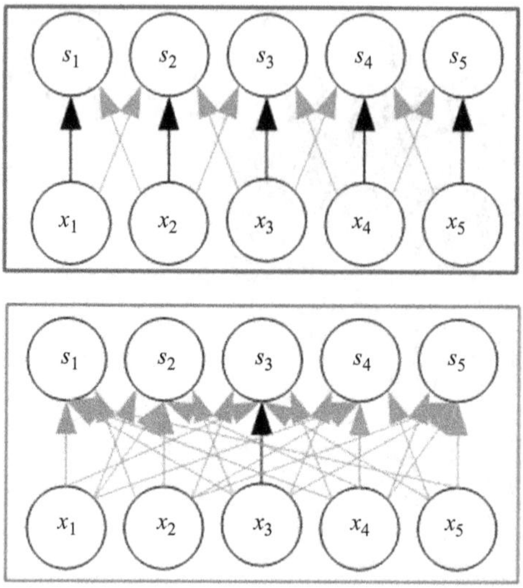

Figure 3.21 Parameter sharing

shown next for LeNet-5, whose first feature map corresponds to a convolutional kernel, which also extracts a feature of the image. Each feature map corresponds to a convolution kernel that extracts a feature of the image. Note here that the final feature map is not the result of convolution, and then a nonlinear activation operation is added. This process is generally referred to as the detector stage.

3.3.2 Pooling

A pooling operation is one in which the values of a region of an image are replaced by all values within that region that are worth computing after the convolution layer has been generated. For example, the maximum pooling operation takes the largest output in a rectangular local region as the output of that region. There are of course other pooling functions, such as average pooling and weighted average pooling. One of the main uses of pooling is that it allows the features obtained after a convolution operation to remain constant over small local translations of the input image. For example, given an image, we are doing image detection and we only care if there is a feature in the image, but its position is less important, so this property of local translation invariance is useful.

Figure 3.22 shows the local translation invariance caused by the maximum pooling. We can see that the input of the lower CL has been shifted right by one neuron, that is, the value of the corresponding neuron in the input layer has changed, but the value of the upper PL has only partially changed. Pooling multiple feature maps can also solve the problem of rotational invariance. The pooling operation allows us to generate another hidden layer of the CNN, the PL. The idea of generating this layer is unambiguous, and the operation is simple. As shown in Figure 3.23, six feature maps are extracted from the original image by applying six convolutional kernels, and then these six feature maps are pooling, also called subsampling, which is the same as the

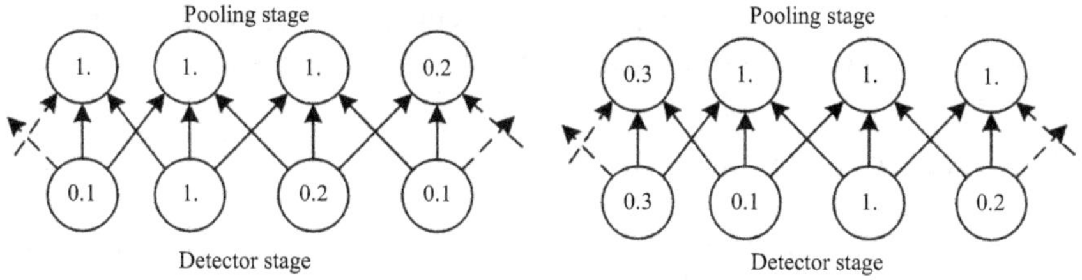

Figure 3.22 Pooling operation

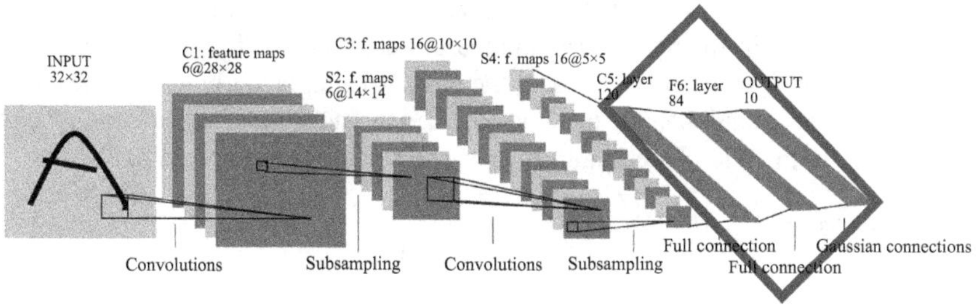

Figure 3.23 Architecture of LeNet-5

Table 3.2 Data distributions and models

Data	Real sample: $x \sim p_{\text{data}}, (x), x \in X$ Latent code: $z \sim p_z(z), z \in Z$ $p_z(z)$ can be a random distribution
Generator $G(z, \theta_g)$	θ_G denotes the parameters of the generator Define implicitly a distribution $p_g(x, \theta_g)$ The generator maps the Z space to X space, that is, G: $Z \to X$
Discriminator $D(x, \theta_d)$	θ_d denotes the parameters of D Decide if the sample is real or fake, that is, D: $X \to R$ $D(x, \theta_d) = [0, 1]$

sparse connectivity and weight sharing mentioned earlier, and the pooling operation will also greatly reduce the parameters of the model.

3.3.3 Fully connected layers

Figure 3.23 indicates architecture of LeNet-5. After several convolutional as well as pooling operations, the network was given a C5 layer, consisting of 120 neurons, and F6, a fully connected layer, which is the same as FNN (Table 3.2). In total, there are $(120 + 1) \times 84$ parameters. Note that here the F6 layer ends up with a sigmoid nonlinear transformation, the result of which is an 84-dimensional binary vector. The final layer consists of Euclidean RBF cells, 1 cell per class, with 84 inputs per cell. This means that each RBF cell calculates the distance between the input vector and the parameter vector, the further the distance, the larger the

RBF. The final loss function is the difference between the neurons in layer F6 and the input weights of a particular RBF neuron, to minimize this difference.

3.4 Recurrent neural network

RNNs are presented for sequence learning. RNN establishes the connection between units as a directed loop. Unlike basic neural networks, where MLP can only map from input data to target vectors. Figure 3.24 indicates the structure of RNN. Before the time series are input into the RNN, the data are converted into 2-D data according to the size of the time window to maintain their sequential nature. W is a common weight, and W and V are the weight matrices of the input and output layers. In addition, when calculating the state S_t at the current time t, the input value x_i is multiplied by the weight U of the input layer and then the result is obtained by multiplying the common weight W, which is summed to be constant, as shown in the following equation:

$$S_t = f(U \times x_t + W \times S_{t-1}) \tag{3.32}$$

$$O_t = f(s_{t-1} \times V) \tag{3.33}$$

Different from classic RNN, the long short-term memory (LSTM) module applies the gating function to its forward and recurrent connections. The following is the complete algorithm for LSTM neural networks (bold symbols indicate vectors or matrices):

$$g^{(t)} = \phi\left(W^{gx}x^{(t)} + W^{gh}h^{(t-1)} + b_g\right) \tag{3.34}$$

$$i^{(t)} = \sigma\left(W^{ix}x^{(t)} + W^{gh}h^{(t-1)} + p_i \odot s^{(t-1)} + b_i\right) \tag{3.35}$$

$$f^{(t)} = \sigma\left(W^{fx}x^{(t)} + W^{fh}h^{(t-1)} + p_f \odot s^{(t-1)} + b_f\right) \tag{3.36}$$

$$s^{(t)} = g^{(t)} \odot i^{(t)} + s^{(t-1)} \odot f^{(t)} \tag{3.37}$$

$$o^{(t)} = \sigma\left(W^{ox}x^{(t)} + W^{oh}h^{(t-1)} + p_o \odot s^{(t)} + b_o\right) \tag{3.38}$$

$$h^{(t)} = \phi\left(s^{(t)}\right) \odot o^{(t)} \tag{3.39}$$

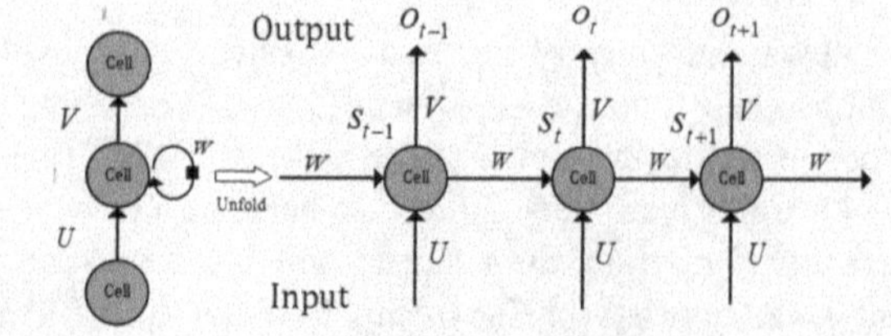

Figure 3.24 The basic structure diagram of RNN

In this algorithm, $g^{(t)}$, $i^{(t)}$, $f^{(t)}$, $s^{(t)}$, $o^{(t)}$, and $h^{(t)}$, respectively, stand for the block inputs, input and forget gates, the cell states, the output gates, and the block outputs of a whole layer at the actual time step. W^{uv} represents for the weight matrix of arbitrary inputs u, v, and b notates bias vectors; h^t is the vector of the output values of the nodes in the current layer with time step t. In addition, p are peephole weight vectors and the operator $\odot$ is used for element-wise multiplication.

Due to the RNN map, the target vector from the entire history of previous inputs and the memory of prior inputs are retained in the internal state of the network. RNN can be trained by backpropagation to accomplish the task of timely supervision. However, the issue of gradient disappearance during the back-propagation of model training hinders the performance of RNN. It means that traditional RNN may not be able to capture long-term dependencies. Therefore, the LSTM algorithm is designed to prevent backpropagation errors from disappearing or exploding. To solve the problem of long-term reliance, forgetting gates are incorporated into LSTM. The use of cell state information may be controlled using these applicable forgetting gates. Because it can capture long-term dependencies, LSTM has the advantage over standard RNN for capturing nonlinear dynamics in time-series sensory input and learning an efficient representation of machine states. Considering the ability of LSTM to capture remote dependencies and nonlinear dynamics in time series data, LSTM has been effectively applied to fault diagnosis of motor drive systems [80,85,86]. Reference [80] presented the RNN-WDCNN, a new dual-path RNN with a larger initial kernel and deeper CNN routes that can operate on raw temporal information. The WDCNN combined the functions of RNN with CNNs to capture remote relationships in time-series data and remove input signals in high-frequency noise. Besides, Reference [85] developed a novel approach named convolutional LSTM (CLSTM) that is developed to handle signal-based FDD in rotating machinery. The developed model improved the efficiency of processing multi-channel input data and learning its spatiotemporal characteristics. The input channels were subjected to sensitivity analysis, and the results suggest that combining these multi-domain characteristics improves the accuracy of the classifier. However, feature engineering on severely unbalanced datasets while finding the optimal hyper-parameters to train the classifier faster without compromising its accuracy is a serious topic. The CLSTM architecture was used to process multi-channel input data and understand its spatiotemporal characteristics more efficiently. The updated CLSTM supported the FDD approach to properly understand the function of data structures and achieve higher accuracy. Reference [86] proposed a deep neural network architecture for processing raw sensory data, called convolutional bidirectional long-term short-term storage network (CBLSTM). The proposed method consisted of one-layered CNN for feature extraction and the size of the input signal is 100×12. Then two hidden layer bidirectional LSTMs were implemented to encode the temporal patterns. Two fully connected layers of size 500×600 were used before feeding the representations to the linear regression layer. The activation function was ReLu. One epoch takes 5 s to train, and each sample takes 0.027 s to test. Compared with RNN, deep RNN, LSTM, and others, the proposed algorithm has better performance. CBLSTM first used CNN to extract stable and information-rich local features from sequential inputs. Then, the temporal information was encoded using a bidirectional LSTM. The LSTM models linear data and record long-term dependencies, while the bidirectional system captures past and future environments. A stacked fully connected layer and a linear regression layer were constructed on top of the bidirectional LSTM to predict the target values. In addition, two RNN networks were applied as encoder and decoder in place of the general fully connected layer to effectively reduce the dimensionality of the time series data combined with the autoencoder of the RNN [87]. Moreover, Reference [53] was used in combination with a variant autoencoder (VAE), and variance and noise are added to make the model generation more realistic. The time-domain vibration signals at three different locations were used as inputs, and two RNN networks were developed as encoder and decoder, respectively, instead of the usual fully connected layer. In addition, the variance and noise were added to make the model generation more realistic, in addition to reconstructing the input data. This method reduced the computational cost. However, when the length of the input sequence was too long, it will cause the information to be diluted during the propagation process, thus decreasing the accuracy.

3.5 Generative adversarial network

A generative adversarial network (GAN) is divided into parts: a generator and a discriminator. GAN training aims to achieve a Nash balance between the generator and discriminator based on the binomial zero-sum game principle. The generator and discriminator in the GAN are differential functions that can be optimized by using any gradient-based approach. The purpose of the generator is to capture the potential distribution hidden in the real sample by using z as input. Therefore, the generator outputs a "fake" sample x_f that is as similar as possible to x_r, thus confusing the discriminator. In addition, the purpose of the discriminator is to distinguish x_f from x_r and to identify the actual sample from the generated sample. Due to the conflicting goals of the generator and discriminator, the two parts of the GAN compete and gradually become more powerful during the training process. Once the GAN is trained correctly, the distribution of the generated samples will match the distribution of the actual samples, which makes it difficult for the discriminator to distinguish the difference between x_f and x_r.

Figure 3.25 indicates the basic principle of GAN. The nonlinear functions of the discriminator and generator approximation are represented as $G(\cdot)$ and $D(\cdot)$,

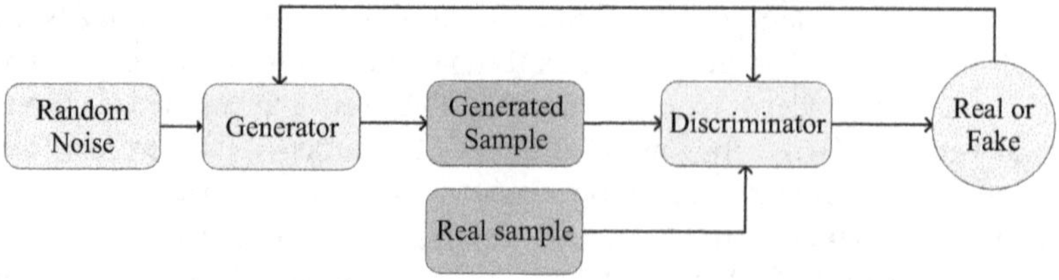

Figure 3.25 The basic structure diagram of GAN

respectively. The distributions of random noise and actual samples are represented as P_{data} and P_n, respectively. Due to the different training objectives of the discriminator and generator, the objective function is defined, respectively, as

$$\{L_G(D, G) = E_{Z\sim P_n}[\text{loglog}(1 - D(G(z)))]\} \tag{3.40}$$

$$\{L_D(D, G) = E_{x\sim P_{\text{data}}}[\text{loglog}D(x)] + E_{Z\sim P_n}[\text{loglog}(1 - D(G(z)))]\} \tag{3.41}$$

where $L_D(D, G)$ and $L_G(D, G)$ represent the objective functions of the discriminator and the generator. Equations (3.11) and (3.12) can be combined into a single-objective function for the overall training operation of GAN, as seen in (3.13):

$$_G\{L_D(D, G) = E_{x\sim P_{\text{data}}}[\text{loglog}D(x)] + E_{Z\sim P_n}[\text{loglog}(1 - D(G(z)))]\} \tag{3.42}$$

Due to their ability to learn deep representations without deep marking of training results (Figure 3.26 and Table 3.2), GANs have received a lot of attention in a wide variety of fields [88,89]. GAN has also been used in mechanical defect diagnosis in recent years [90,91]. The data imbalance between different machine health conditions can be addressed [92,93] and unsupervised fault classification can be achieved by incorporating GAN into certain DL algorithms [91]. In general, ML- and DL-based FDDs share a universal drawback, which is the models require a substantial quantity of training data to learn the inherent patterns of normal and faulty data in a customized manner. However, for various reasons, it is hard to acquire actual operational data corresponding to different health conditions: (1) low frequency of failures, (2) inserting faults into rotating equipment is expensive and dangerous, (3) labeling data are highly time-consuming even if a large amount of data are available. Therefore, GAN is a method that requires only a small amount of sample data to obtain highly accurate training results. GAN can generate sample ML algorithms in an unsupervised manner. Due to the adversarial generator and discriminator, the GAN will learn samples adaptively for training. When both the generator and discriminator have completed training, the output of the generator will produce samples close to the actual samples even if the input data are random noise. On a variety of image datasets, GAN variants such as DC-GA and AC-GAN have demonstrated strong generative efficiency. Reference [94] applied the GAN method

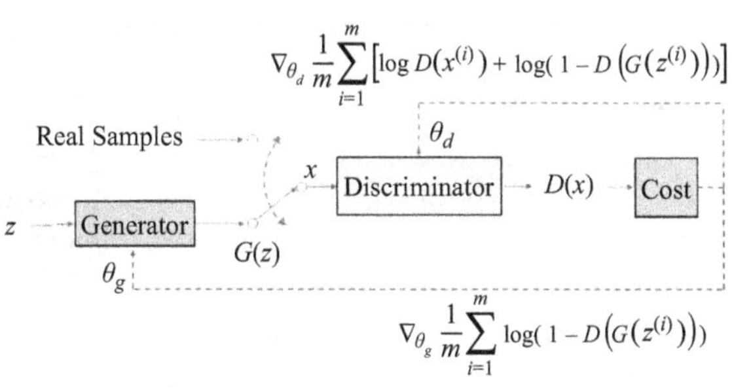

Figure 3.26 GAN structure and training process

to the FDD of a motor instead of other traditional resampling methods. To increase the imbalance data of the motor, the GAN-based method generated more realistic samples and improved the accuracy of fault diagnosis compared to use other resampling methods. Owing to a complete lack of anomalous samples, single GAN-based fault diagnosis models do not recognize the case of new types of faults. Using stacked noise reduction autoencoders and GAN, Reference [95] proposed a fault diagnosis process (SDAE-GAN) that enabled the discriminator to determine the sample fault type and whether the input data are from a real data distribution. However, learning various data distributions simultaneously is a challenge. Furthermore, the proposed method can eliminate the interaction between the intrinsic distributions of different fault patterns during the training process. Therefore, it has stronger generalization capability compared with a single-fault diagnosis method based on GAN. This method is of great significance for solving the problem of difficulty in obtaining fault data in practice. Although FFT requires less expertise than other manual feature extraction methods, it still requires the use of signal processing, and the proposed method cannot be considered an end-to-end fault diagnosis method. Reference [96] studied multiple GANs to understand the data distribution for each health condition and then developed a semi-supervised approach to improve the feature extraction capability of each GAN with better generalization capability compared to a single GAN-based approach. The generator and discriminator have a symmetrical structure that has one input and output layer and two hidden layers. The size of the input layer of the generator is determined by the size of the random noise vector which is 128×128. ReLu is applied for the activation function of hidden layers, while SIGMOD is used in the output layer. Besides, only 200 samples in each dataset are used for training the model. The performance of testing accuracy is greater than 95% and the standard deviations are below 1.2%.

3.6 Deep belief network

In contrast to traditional neural network discriminative models, deep belief networks (DBNs) are probabilistic generative models in which the generative model creates a joint distribution between the observed data and the labels. Layer-by-layer training allows better initial weights to be assigned to the entire network so that the network can be fine-tuned to achieve the best solution. DBNs consist of multiple layers of constrained Boltzmann machines (RBMs), a typical neural network as shown in Figure 3.12. These networks are "constrained" into visible and hidden layers, with connections between the layers, but there are no connections between the internal units. The hidden layer units are learned to recognize the relevance of the higher order data displayed in the visible layer. Through the energy function $E(v,h,\theta)$,the relation weights describe a probability distribution over the joint condition of the visible and hidden cells (v,h):

$$E(\mathrm{v},\mathrm{h},\theta) = -\sum_{j=1}^{m} b_j v_j - \sum_{i=1}^{n} c_i h_i - \sum_{i=1}^{n}\sum_{j=1}^{m} v_j w_{i,j} h_i \tag{3.43}$$

The probability distribution of each possible visible and hidden vector pair can be defined by the following energy function similar to that of a general Boltzmann machine:

$$p(v, h;\theta) = \frac{1}{Z(\theta)} \exp \exp(-E(v, h;\theta)) \tag{3.44}$$

where $Z(\theta) = \sum^{v,h} \exp(-E(v, h))$. The conditional probabilities of hidden and visible cells are given by

$$p(v;\theta) = \frac{1}{\left[1 + \exp\left[-c_i - \sum_{j=1}^{m} v_j w_{i,j}\right]\right]} \tag{3.45}$$

$$p(h;\theta) = \frac{1}{\left[1 + \exp\left[-b_i - \sum_{i=1}^{n} v_j w_{i,j}\right]\right]} \tag{3.46}$$

The RBM model with binary cells can be learned by a negative log-likelihood gradient as follows:

$$\Delta w_{i,j} = \eta(\Delta w_{i,j} = \eta\langle v_j h_i\rangle_{p(h|,v)} - \langle v_j h_i\rangle_{\text{recon}}) \tag{3.47}$$

$$\Delta b_j = (\langle v_j\rangle_{p(h|v)} - \langle v_j\rangle_{\text{recon}}) \tag{3.48}$$

$$\Delta c_i = (\langle h_i\rangle_{p(h|v)} - \langle h_i\rangle_{\text{recon}}) \tag{3.49}$$

where η is the learning rate and $\langle \cdot \rangle_{p(h|v)}$ is the expectation value, which is relative to the conditional distribution $p(h|v)$ reconstruction of the distribution of the model. Superposition in the forward direction RBM learning may be used to learn weights in an unsupervised manner, which is referred to as supervised initialization of the learning parameters. It is the same as having a priori knowledge of the supervised learning input data.

Some DBN-based FDD methods are carried out in the motor drive system (Figure 3.27). A novel hierarchical diagnosis network (HDN) rolling carrying automatic diagnosis method, consisting of two layers of DBN, was proposed in [97]. WPT was used to provide representative features to deal with the non-smoothness of

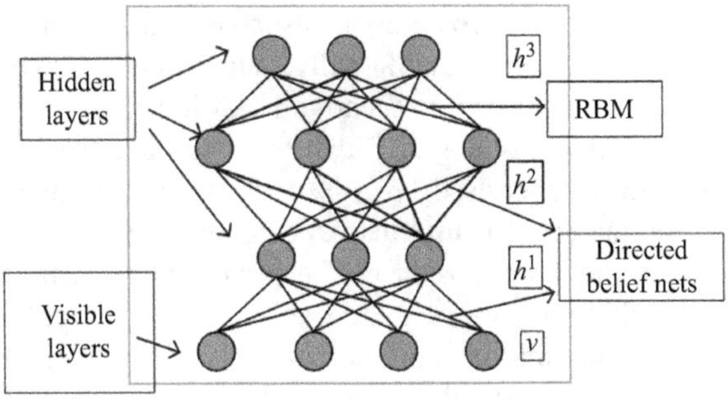

Figure 3.27 The basic structure diagram of DBN

the fault vibration signal. Determination of fault location and its magnitude level can be achieved by HDN, and weak points were identified in this way to avoid device performance loss and providing information for the reliability of the designed configuration. Reference [98] proposed an improved convolutional deep confidence network (CDBN) with compressive sensing (CS). The structure of the proposed algorithm was divided into one Gaussian input layer, 6 convolutions hidden layers, 6 PLs connected with 12 convolutions hidden layers, and 12 PLs. The output was fed into a SoftMax classifier for fault classification. The compressed vibration signals as input data and the two cases that were considered for verification include single fault and compound faults. The performance of testing accuracy was 94.80% and the standard deviation was 0.53. To improve the analysis efficiency, CS was used to reduce the amount of vibration data. Then, a new CDBN model with Gaussian visible units was developed to enhance the functional learning capability of compressed data. Furthermore, the constructed deep model generalization efficiency was improved using the exponential moving average (EMA) technique. A discriminative deep belief network (ACO-DDBN) based on ant colony optimization was presented in [99]. The proposed method consists of one input layer, two hidden layers, and one output layer. The input signal is a vibration signal with 1,024 data points. It takes 1,592.5 s for training the model. After the training process, the training accuracy and the testing accuracy are 94.7% and 91.2%, respectively. The RBM greedy layer-by-layer learning algorithm can effectively pre-train unsupervised models regardless of the amount of available training data. However, the performance of deep belief networks can be affected by their parameters. Since ants find the best choice of parameters throughout the search process, ACO is suitable for selecting parameters. Therefore, the parameters of the model are obtained by using the ACO method. The composition of the DDBN model can be computed automatically without the optimization process and prior knowledge to improve efficiency.

3.7 Autoencoder

The autoencoder neural network is an unsupervised learning algorithm that makes the target value equal to the input value by using a backpropagation algorithm. Figure 3.28 illustrates the basic structure diagram of AE. An autoencoder is a neural network that consists of two components: an encoder function and a decoder that generates reconfigurations. Traditionally, autoencoders have been used for dimensionality reduction or functional learning. An improved prototype structure of the autoencoder has resulted in a DAE that first complete the pre-training of the hidden layer using an unsupervised layer-by-layer greedy training algorithm and then uses BP to tune the entire neural network through systematic parameter optimization. This algorithm greatly reduces the performance metrics of the neural network and effectively improves the undesirable situation where the BP algorithm tends to fall into local minima. In short, compared with the original AE, DAE can increase depth, improve learning ability, and help pre-training. SAE only changes one thing: The input data are fed into the model with noise, and then the original

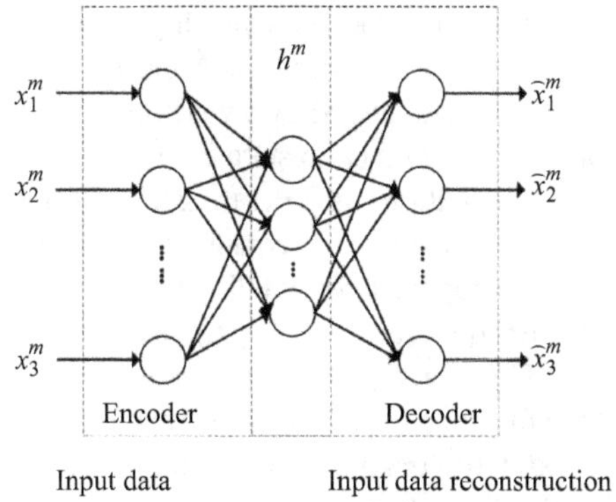

Figure 3.28 The basic structure diagram of AE

image without noise can be restored. Through this method, a better robust representation can be extracted.

The dimensionality reduction includes the encoder network. The function reduces the initial high-dimensional data to a low-dimensional representative. The decoder network is the opposite of the encoder network and is part of the restoration process. The function returns the studied low-dimensional features to the high-dimensional data. The equations are shown as follows:

$$h_i = f_{\theta_1}(W_1 x_i + b_1) \tag{3.50}$$

$$\widehat{x}_i = g_{\theta_2}(W_2 h_i + b_2) \tag{3.51}$$

where W and b are the weights and biases of the network.

The purpose of the autoencoder model is to learn useful hidden representations by minimizing the reconstruction error. Therefore, the parameter sets θ_1 and θ_2 can be optimized by minimizing the reconstruction error L_{rec}, while giving n training samples:

$$L_{\text{rec}} = \frac{1}{n}\sum\nolimits_{i=1}^{n} \|x_i - \widehat{x}_i\|^2 \tag{3.52}$$

$$\widehat{\theta}_1, \widehat{\theta}_2 = \arg \operatorname{arg} L_{\text{rec}}(\theta_1, \theta_2) \tag{3.53}$$

where $\widehat{\theta}_1$ and $\widehat{\theta}_2$ denote the optimal values of θ_1 and θ_2, respectively. The stochastic gradient descent (SGD) algorithm can be used to solve this optimization problem. The equations are shown as follows:

$$\widehat{\theta}_1 \leftarrow \theta_1 - \lambda \frac{\partial L_{\text{rec}}}{\partial \theta_1} \tag{3.54}$$

$$\widehat{\theta}_2 \leftarrow \theta_2 - \lambda \frac{\partial L_{\text{rec}}}{\partial \theta_2} \tag{3.55}$$

where λ is the learning rate.

A monitoring method based on stacked denoising autoencoders was proposed in [100]. The stacked denoising autoencoder-based approach can effectively extract robust features from corrupted data and shows well potential in the field of process monitoring. It is suitable for specific health state identification for signals containing environmental noise and fluctuations in operating conditions [101,102]. A novel method namely local connection network constructed by normalized sparse autoencoder (NSAE-LCN) was proposed for intelligent fault diagnosis [103]. In this approach, the LCN first learned various meaningful features from the vibration signal using NSAE. Then, the LCN generated displacement-invariant features to classify the health condition of the machine based on the learned features. In [104,105], a new DL algorithm for bearing fault diagnosis was proposed. A combination of discriminative and structural information between different fault conditions in a deep autoencoder model was applied. The method can study the association structure and structural relationship information among multiple fault states, which helped to improve the stability of the deep neural network. To effectively assess the health of a motor, various sensors were installed in different locations to obtain more fault signals. However, due to different sensor arrangements and environmental disturbances, the acquired signals may change, which led to different diagnostic results. To improve the reliability of diagnosis, a signal fusion technique was proposed in [106]. The proposed method combined SAE and DBN. Data features were extracted from sensors and fed into SAE for feature fusion. The SAE has two hidden layers with the same structure and parameters, while DBN has three layers. The training data were vibration signals of under inner race fault and outer race fault with 1,260 samples. The classification accuracy was up to 97.82%. For feature fusion, time and frequency domain features from various sensor signals are extracted and fed into several two-layered SAE neural networks. The fused feature vectors are used to machine health indicators. Reference [107] proposed a simple DL clustering method, inspired by [43], which applied manifold learning to off-the-shelf embeddings to find an alternative model for clustering networks and simply combined the manifold learning the input layer was equal to the size of input data. The ReLu was applied for the activation function, while the Adam optimizer was implemented. Bearing fault data were used for experimental validation. The dataset contained 10 types of bearing health conditions with 10,000 samples. The average classification accuracy was up to 98.99%. Instead of complex clustering networks, the shallow clustering algorithm presents a new deep clustering method for E2LMC, which is based on autoencoder embedding local stream shape learning for unsupervised bearing fault diagnosis.

Autoencoder is an unsupervised learning algorithm for learning a valid representation from the input data. It consists of two components: encoder and decoder. The encoder maps the input data to a low-dimensional representation in the latent space, while the decoder maps that low-dimensional representation back to the original input space, reconstructing the original data as closely as possible. The training process of the autoencoder aims to minimize the reconstruction error, that is, the difference between the input data and the output of the decoder. By training in this way, the autoencoder learns important features of the data and generates encoder–decoder pairs that can reconstruct the original data.

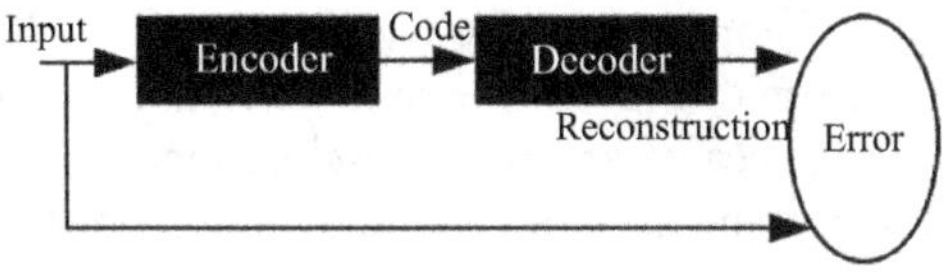

Figure 3.29 Diagram of autoencoder

Input: the input of the data;
Encoder: encoder;
Code: a representation of the input;
Decoder: decoder;
Reconstruction: reconstruction of the input;
Error: the error of the reconstructed data and input.

As shown in Figure 3.29, a common type of autoencoder is a feedforward autoencoder based on a fully connected neural network. It consists of an input layer, a hidden layer (part of the encoder), and an output layer (part of the decoder). The hidden layer typically has a lower dimensionality than the input and output layers, which forces the autoencoder to learn a compressed representation of the data.

An autoencoder is a neural network that reproduces the input signal as closely as possible. The auto encoder must capture the most important factors that can represent the input data.

3.7.1 Probabilistic interpretation of reconstruction errors

Objective function:

$$L = -\log P(x|g(f(x))) \tag{3.56}$$

The choice of the loss function depends on the type of input data: If the input is a real, unbounded value, the loss function uses squared error; if the input is a bit vector, cross-entropy is a more suitable loss function. We can think of training a decoding function as equivalent to estimating the conditional distribution $P(x|h)$; by the same token, the encoding function can be understood as a conditional distribution $Q(h|x)$, rather than just a "dry" function equation. The cross-entropy of the p and q distributions is the sum of the information entropy of the p distribution and the DL scatter of p and q:

$$H(p,q) = E_p[-\log q] = H(p) + D_{KL}(p||q) \tag{3.57}$$

3.7.2 Sparse autoencoder

$$\text{Input}: \ \text{X code}: h = W^T X \tag{3.58}$$

$$\text{Loss:} \ L(X;W) = ||Wh - X||^2 + \lambda\Sigma_j|h_j| \tag{3.59}$$

If we add the regularity restriction of $L1$ to autoencoder ($L1$ essentially constrains most of the nodes in each layer to be 0 and only a few not to be 0, which is where the Sparse name comes from), we get the SAE method. As shown in Figure 3.30 earlier it is a matter of limiting the expression code obtained each time to be as sparse as possible. This is because sparse expressions tend to be more effective than others (the human brain seems to work the same way, where a certain input just stimulates some neurons and most of the others are inhibited).

Table 3.3 summarizes the DL methods for fault classification. The merits and demerits of DL are presented as follows:

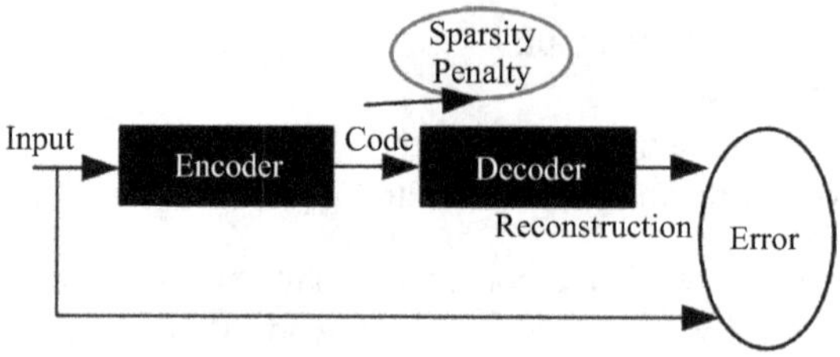

Figure 3.30 Diagram of sparse autoencoder

Table 3.3 Highlight of deep learning

Reference	Theme	Principle	Highlight
[77]	DCNN	• Multiple hidden layers • Learn hierarchical representations	• Deep mining signal features • To require large training dataset
[74]	CNN	• 1-D convolutional neural network • Scalar multiplication and addition • Raw data input	• To decrease computing cost • High efficiency • Insufficient classification accuracy
[75]	Sparse-filtering CNN	• Unsupervised two-layered neural network • SoftMax regression	• Adaptive learning feature • No manual labeling required
[76]	CCA KLD CNN	• Moving average technique • Fault matrix input	• Statistics are more sensitive to initial failures • To improve computational efficiency • No prior knowledge required
[78]	CNN RNN	• Vibration signal input	• Efficient and time-saving
[79]	CNN ELM	• Weighted orthogonality constraint • Randomly generate hidden weights	• To reduce training complexity • To reduce manual intervention • To improve generalization performance

(Continues)

Table 3.3 (*Continued*)

Reference	Theme	Principle	Highlight
[80]	RNN-WDCNN	• Vibration data input • Domain adaptation and noise suppression	• No feature extraction required • High Robustness
[81]	WDCNN	• Raw vibration signal input • Adaptive batch normalization	• To improve network accuracy • No preprocessing required
[82]	TICNN	• End-to-end learning • 1-D structure input • Direct denoising	• No manual feature extraction process required • High robustness • High accuracy
[83]	SVM-CNN	• Unsupervised fault detection • Online multi-fault classification	• High generalization ability • Robustness
[84,108]	CNN	• 2-D image conversion • Eliminate the impact of manual features • Easier handling of big data	• Time-consuming training • High hardware conditions • Great training capabilities
[85]	FFT CWT CLSTM	• Multi-channel array input • Input length sensitivity analysis	• Efficient learning of spatiotemporal features • To enhance classification performance
[86,109]	LSTM CBLSTM	• Original sensing signal composition • Revealing fault characteristics	• Abstraction and deep learning ability • No expert knowledge required
[87,53]	AE-RNN	• Time series data dimensionality reduction • Raw vibration signal input	• To reduce dimensionality effective feature • Extraction • To improve classification efficiency
[110]	SGU RNN	• Single-gate joint recurrent neural network • Wavelet packet decomposition	• To reduce network parameters • To reduce computation time
[94,96, 111]	GAN	• Unsupervised sample generation • Adaptively learning samples • Generated sample input	• Strong generalization capability • Non-original feature signal input • Not an end-to-end structure
[95]	SDAE-GAN	• Estimating the probability distribution	• Automatic extraction of effective fault • Characteristics

(Continues)

Table 3.3 (*Continued*)

Reference	Theme	Principle	Highlight
		• Extended diagnostic sample	• Stronger noise immunity • Improved generalization capability
[57]	DAAN	• Feature compression and visualization • Vibration signal input	• Effectively learn domain-invariant features. • Diagnosis under different operating conditions
[98]	CDBN CS	• Reduced vibration data • Layered generation model	• No manual feature extraction required • Great generalizability
[99]	ACO DDBN	• Layer-by-layer unsupervised pretraining • Search for the optimal number of neurons • Learning rate of the hidden layer	• Accurate determination of model parameters • To increase the computational burden
[54,106]	SAE DBN	• Feature fusion • Unsupervised learning weights	• Insensitive to training samples • High classification accuracy
[97]	HBN	• Two-layered DBN • Vibration signal input	• High reliability • To overcome overlap problem caused by noise
[52,101]	SDA	• Health status division • Greedy training	• Adaptive mining of salient fault characteristics • High diagnostic accuracy • Strong robustness
[103,105]	NSAE LCN	• Raw vibration signal input • Obtain translation invariant features	• To generate displacement invariant • characteristics • Effectively identify the mechanical health • condition
[104]	AE	• Association matrix	• To dig complex information • High diagnostic accuracy

Merits:

1. High learning ability: DL performs well and has a great learning ability.
2. Strong adaptability: DL contains complex neural network layers that can be mapped to any function to solve complicated classification problems.
3. Well portability: Many frameworks can be used.

Demerits:

1. Computationally intensive.
2. DL requires large data and computing power so that the cost is extremely high.
3. High hardware requirements.
4. Complex model design.

3.8 Other AI-based approaches

In addition to the traditional ML and DL introduced earlier, some other popular artificial intelligence-based methods are also reviewed in this section, including hybrid algorithms combining fuzzy logic (FL) and artificial neural network methods (ANFIS), transfer learning (TL), compressed sensing (CS), infrared thermography (IRT) and temperature estimation strategy, genetic algorithms, and others. Table 8.1 (Chapter 8 of this book) summarizes the characteristics of each AI-based method.

3.8.1 Adaptive neuro-fuzzy system (ANFIS)-based approach

ANNs have proven to be a reliable technique for diagnosing motion conditions with high learning capabilities. However, ANNs are not interpretable and cannot explain specific decisions to the user in a human-understandable form. Another technique used for FDD is FL. It can imitate human knowledge according to clear and understandable linguistic terms and then convert the linguistic and heuristic terms into complex machine computed values through fuzzy rules and auxiliary functions. The initial parameters and auxiliary functions of if-then rules are usually prepared by experts. Therefore, FL needs to be fine-tuned to obtain an acceptable rule base and to optimize the parameters for the available data. The integration of these two methods can solve only one problem from FL or ANN. This method has been applied to motor fault diagnosis [112]. Adaptive neuro-fuzzy inference system (ANFIS) [113] is a special neuro-fuzzy classifier approach that combines the adaptive capabilities of ANNs with an FL qualitative approach. It has been successfully implemented for automatic FDD of induction motors [114]. In recent years, ANFIS and its variations with other methods have been widely developed as fault classification techniques. In terms of bearing fault diagnosis, ANFIS with genetic algorithm [115] and ANFIS with wavelet transform [116] are two examples of combined algorithms. ANFIS has been used to classify faults in induction motors with variable drive speeds [117]. If the measurement data are large and include redundant noise, the accuracy of the output will be greatly reduced when the data are fed directly into the classifier. Feature extraction and selection can minimize the dimensionality of the data by selecting the basic features, which refer to the transition of the current features to the lower dimensional space. However, when feature extraction is completed, each feature set contains several redundant or unnecessary features, as well as significant features in the feature space. Therefore, a feature selection process is required to select the minimum features that define the system state from the entire feature set. As a feature extraction method, decision trees were employed in [118], which was a process for removing redundant features from data to reduce the quantity of data required for efficient learning, classification accuracy, a compact and simply understood knowledge base, and rapid computation. For fault diagnosis of induction motors, it integrated CART and ANFIS-adaptive techniques.

Since insufficient accuracy of the measurement equipment, model errors, or the influence of environmental conditions of the measurement process leads to noise in the signal that usually cannot be avoided, noise filtering mechanisms are important to improve the ability to preprocess the measured data for analysis and extraction of valuable information. Singular spectrum analysis (SSA) and sparse filtering (SSA/SF) are promising tools, and the effectiveness of FL depends heavily on the accuracy of the fuzzy set, which involves logical relationships between fuzzy rules and relational functions input and output. Reference [119] proposed an ANFIS based on SSA/SF bearing fault detection method, which was combined with an online detection method. Information was extracted from the measurement data with noise by SSA and SF, and a database with tagged data was created. The ANFIS parameters were optimized by the tagged data and then faults were detected by the ANFIS comparator. In [59], a higher order particle filter was developed to make predictions based on an m-order HMM that integrates ANFIS and modeling noise. Advanced particle filters were applied to system status modeling. By combining ANFIS and process noise, a high-order HMM was formed that can be used to describe the fault process. Due to the dynamics of the system, an online model adaptive scheme for fault propagation was required. ANFIS was trained with available conditional data to build ANFIS for laminar imaging trend models. A new classification method based on time–frequency representation (TFR) and criteria-based decision-making was proposed in [128]. The fuzzy Doppler delay plane was used as a feature extraction space to reduce the computational effort. This is an error probability model based on a statistical approach for selecting the optimal number of features to be extracted from the streamwise plane and then using the capability of a neural network to learn a nonlinear functional relationship between the input and output. The model is designed to maximize the separability among different classes.

Fuzzy systems and neural network control are two important branches within the field of intelligent control. Fuzzy systems are a class of systems designed to emulate the human approach to FL thinking, an approach that itself clearly states that the system allows for qualitative knowledge in its working process. On the other hand, the learning ability and fault tolerance that neural networks exhibit during the computational processing of information comes from the structural characteristics of the network itself. Fuzzy neural networks are a new technology that combines the powerful structural knowledge representation capability of FL reasoning with the powerful self-learning capability of neural networks and is the result of an organic combination of FL reasoning and neural networks.

Fuzzy control systems are based on several fuzzy linguistic empirical rules that have been summarized by people to describe the interrelationships between various factors, and these empirical rules are translated into simple numerical operations to allow machines to implement these rules concretely instead of people in front of the corresponding problems.

In the general design of fuzzy systems, rules are expressed linguistically by professionals who have extensive experience with the problem being solved.

3.8.2 Fuzzy control combined with neural networks

As shown in Figure 3.31, the model uses fuzzy control as the main body and applies a neuronal network to implement the decision-making process of fuzzy control, using the fuzzy control method as the "sample" and training the neural network offline. The "samples" are the "teachers" who learn. After all the samples are learned, the neuronal network is a smart and flexible fuzzy rule table with self-learning and self-adaptive functions.

3.8.3 Fuzzy neural system—neural network fuzzification, which is still essentially an ANN

As shown in Figure 3.32, the model takes a neural network as the main body, divides the input space into several different types of fuzzy inference combinations, makes an FL judgment on the system first, and uses the fuzzy controller output as the input to the neuronal network. The latter has self-learning intelligent control characteristics.

3.8.4 Fuzzy-neural hybrid system—an organic combination of the two

As shown in Figure 3.33, this model is based on the different nature of the input quantity respectively by the neural network and fuzzy control to directly process the input information and act on the control object and better play the respective control characteristics.

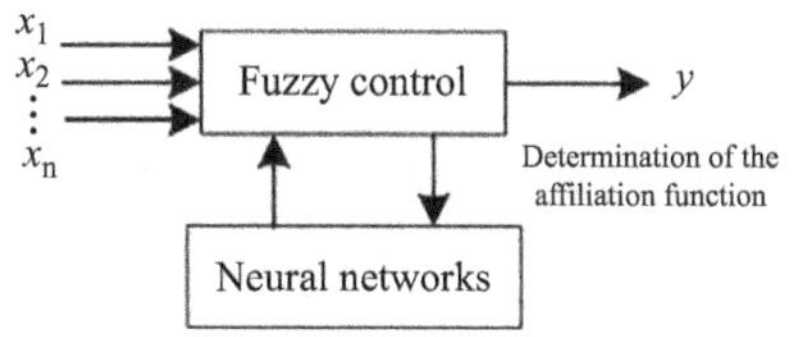

Figure 3.31 Combination of fuzzy control and neural network

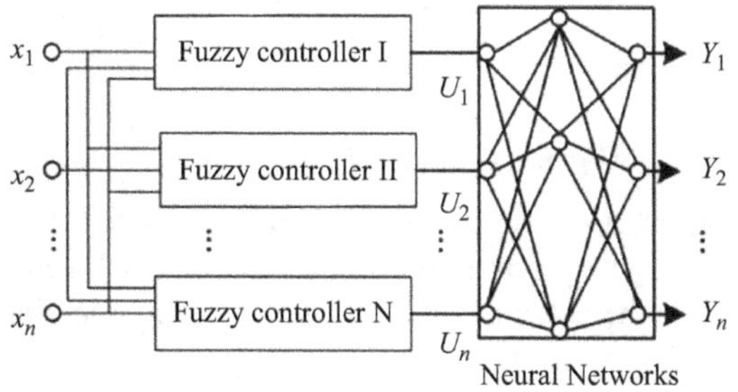

Figure 3.32 Neural network fuzzification

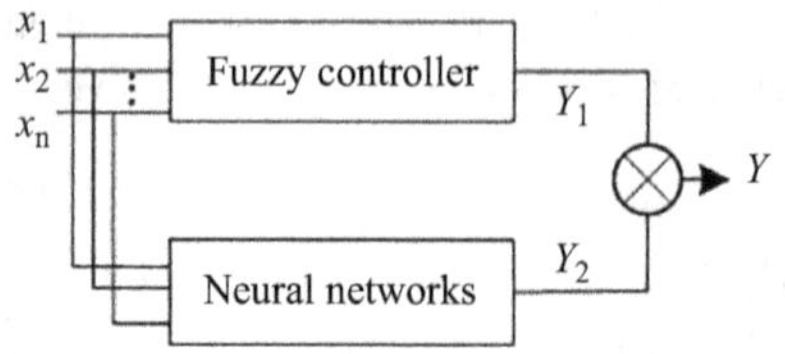

Figure 3.33 Fuzzy-neural hybrid system

3.8.4.1 Advantages of combining neural networks and fuzzy control

1. The structure and weights of the network have a clear physical meaning, and the structure of the network and the initialization of the weights have a theoretical basis to avoid the network falling into a local optimum.
2. The learning ability of the neural network can be used to adjust the control rules and fuzzification methods of fuzzy control, making the fuzzy control have a certain self-adaptive ability.
3. The fuzzy neural network combines qualitative knowledge expression and quantitative numerical operation, which has a good control effect.

3.8.5 Fuzzy association memory

In a fuzzy control rule, the fuzzy rule $A_i \rightarrow B_i$ represents a basis for reasoning, which is an abbreviated form of the implication "If A_i, then B_i."

$A_i \rightarrow B_i$ can be described as a mapping relation in the input–output space of fuzzy control:

$$R_i = (A_i \rightarrow B_i) = A_i \times B_i = A^T B \tag{3.60}$$

The set of all fuzzy control rules implies a fuzzy mapping relationship that can be described as

$$R = \bigcup_i R_i = \bigcup_i A_i \times B_i \tag{3.61}$$

For a fuzzy input A, fuzzy inference can be performed based on a fuzzy rule base to obtain its corresponding fuzzy output as

$$B = A \circ R = A \bigcup_i R_i = \bigcup_i A \circ R_i \tag{3.62}$$

3.8.6 Structure of the fuzzy association memory

A premise A acting on m bases of reasoning produces m sub-conclusions simultaneously (Figure 3.34). These m sub-conclusions are combined to give the final output of the FAM as the conclusion. Here, the so-called combination process is a normalized weighted summation:

$$B = W_1 B_1' + W_2 B_2' + \cdots + W_m B_m' \tag{3.63}$$

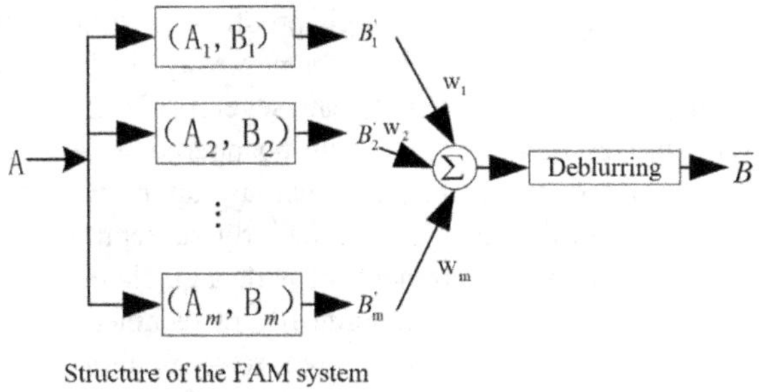

Figure 3.34 The diagram of FAM system

During the control process, the impact of each rule on the output can be changed by changing the weight factor W_i according to the situation, thus achieving the goal of adjusting the performance of the fuzzy control.

3.8.7 Compressed sensing

The increasing demand for information about complex mechanical systems in recent years has promoted the use of high-dimensional information. To examine health conditions comprehensively, monitoring systems based on traditional sampling theorems are commonly employed. Multiple sensors acquire a large amount of data at high sampling rates over a long operating cycle that puts a burden not only on the hardware but also on the data storage. In the era of big data driving motor health monitoring, how to extract effective information from a large amount of data is an important topic, and CS has attracted widespread attention. This method theoretically obtains all the information contained in the original signal and achieves the reduction of the signal in size. It frees the data from Nyquist theory and enables compressive acquisition by nonlinear projection in the transform space. Thus, the sampled data are greatly reduced and contain all the information. Fault diagnosis is achieved based on the measured data in the compressed domain, which is supported by the depth domain. A new intelligent FDD combining CS and DNN was proposed in [120]. Random projections in the transform domain were used as compressed samples to perform sample compression, and then a stacked autoencoder network based on SAE was built on mining fault information using a deep network architecture. In addition, the detection of multiple hybrid faults is an essential topic. Reference [121] presented a diagnostic method for tracking and analyzing the frequency of multiple and combined faults in IM, depending on the measurement and optimization of a high-dimensional hybrid feature set. The proposed method can detect the possibility of multiple and combined faults occurring simultaneously.

3.8.8 Infrared thermography and temperature estimation strategy

When constructing an electric motor or selecting its control approach, the significant thermal stress on possibly defective components of the motor must be taken into consideration. Competitive pressure and high production costs, particularly in the automotive industry, lead engineers to look for new ways to increase the safety of embedded materials. Because overheating can severely damage motors, accurate temperature information must be provided during operation. Among the typically critical components sensitive to overheating, such as stator end windings and bearings, permanent magnets in the rotor are particularly susceptible to damage to the motor. Although sensor-based measurements provide a quick and accurate picture of the machine's thermal state, it is not economically or technically feasible to assess rotor temperature in this way. In particular, direct rotor monitoring techniques, such as IRT [122,123] or classical thermocouples with slip rings on the shaft [124], have not yet entered industrial mass production. Therefore, based on this model, the study focuses on estimating the rotor and permanent magnet temperatures. In contrast to physically driven estimation methods, ML models will differ from any classical approximation of the underlying thermal theory [125]. Figure 3.14 shows the idea of fitting the ML model to the collected test bench data and eventually informing any controller. The more accurate the thermal state information obtained by the control system, the better the ability to monitor critical operations and apply derating power. In addition, suppose rich datasets can be recorded on the test bench or in production in the automotive industry, in that case, engineers can rely on them to model simpler temperature estimators than deep neural networks. The use of domain-specific expertise and instrumentation specifications can be avoided. In addition, certain classical supervised learning algorithms can handle the temperature of essential components in PMSM in real time, as demonstrated by extensive benchmark analysis of many different benchmarks. Alternatively, IRT can be used to detect bearing failures in rotating machinery in a nondestructive, noncontact manner. However, performance is limited by the minuscule amount of information and string noise in IR thermal images. To address this problem, Reference [126] developed a 2-D-DWT-based IRT technique for identifying different bearing faults in IM, such as inner and outer ring cracks and insufficient lubrication, using a 2-D-DWT-based IRT method. PCA was used to decrease the dimensionality of the recovered features to obtain the best feature selection and then a mar distance procedure was used to rank the selected features in order of importance. For fault classification and performance evaluation, these chosen characteristics were input into complex decision trees, linear discriminant analysis, and SVM (Figure 3.35).

3.8.9 Transfer learning

Fault signal processing includes the separation and extraction of fault features for later use in predictive models. Fault discrimination identifies fault types by testing the fault features obtained through signal processing using a complete training model. Typically, the number of fault samples is usually smaller than the number of regular samples, and the problem of data class imbalance will lead to the overfitting of many data-sensitive classification models. The data imbalance problem can be solved by weighting cost, but for practical motor fault diagnosis, it is difficult to determine the weighting parameters. Therefore, obtaining as much fault data as possible from data

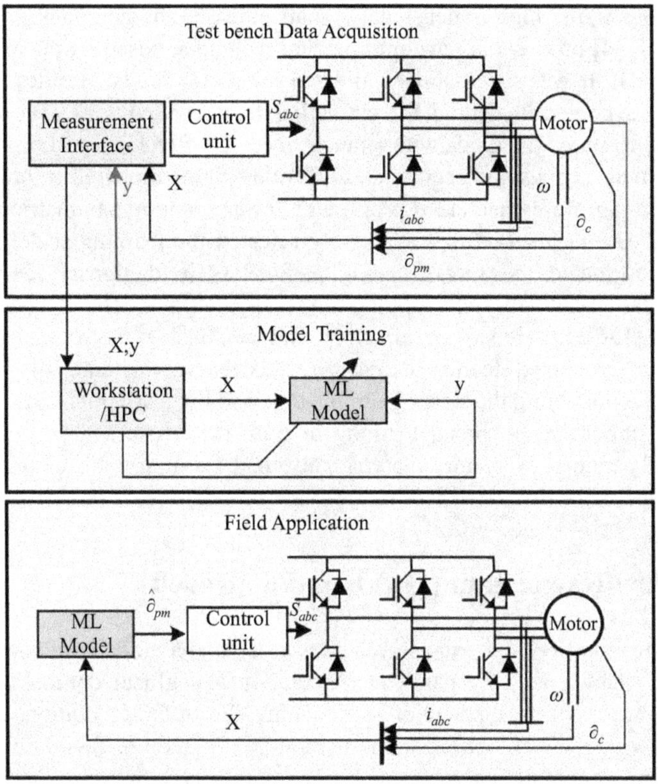

Figure 3.35 The entire process test bench

sources is a major issue. Simulation of fault scenarios is the easiest way to obtain fault characterization results. However, it is impractical for mechanical parts of motor equipment to intentionally damage them to achieve fault characteristics. Therefore, it is a challenge to make full use of historical data and ensure the accuracy of the new task model. Reference [127] proposed a method to solve the problem of degradation of classification accuracy due to different fault data caused by feature transfer. In feature learning, transfer component analysis (TCA) is a classical approach to solve adaptive problems. Data in both source and target domains are mapped to a high-dimensional replica kernel Hilbert space. The data distance between sources is minimized, while interior features are preserved. The TCA data feature extraction method is used to discover standard features in both the source and target domains. Through feature learning, a fault diagnosis model under historical operating conditions can be built to address the problem of insufficient data effectively.

3.8.10 Artificial ant system (AAS) and dictionary learning

A uniform sampling window samples the signal, while the more petite frame width and smaller size are insufficient to accommodate the input signal [133].

Simultaneously, the high-dimensional signal increases the computational burden. Reference [134] proposed a new optimization technique based on the artificial ant system (AAS). It was successfully validated for motor faults. Besides, dictionary learning is a powerful method for extracting feature conditions. The multiscale dictionary learning method with transformed coefficients performs well in extracting fault signals and requires less learning time compared to ordinary dictionary learning. Reference [135] proposed a faster adaptive parametric multiscale dictionary learning method that adaptively selected the learning scale and simultaneously estimated the sparse coding parameters in dictionary learning. The technique has a relatively short running time and great fault detection performance. Reference [136] developed an effective and responsive motor fault diagnosis method that combined feature extraction, and feature incremental generalized learning after collecting the starting sample data and processing it. These processed data were imported into broad learning to train the network. The network was continuously trained by feature incremental broad learning until the test accuracy was satisfactory.

3.9 Genetic algorithm (GA)-based approach

Genetic algorithms (GA) are the search ones based on natural selection and genetic principles, characterized by parallel computation and global optimality, as their derivation process is a method of approaching the optimal solution. Moreover, compared with general optimization methods, GAs require less information to achieve optimal control. Therefore, GAs are often used to optimize the parameters and structure of neural networks or FL systems. Reference [28] presented a diagnostic method based on the envelope, FFT, and backpropagation GA-ANN algorithm for rolling bearing faults to determine the NN structure accurately. The proposed GA-ANN combination strategy can be used to assist experts in designing appropriate, fast, and accurate ANN structures for specific diagnostic problems. Besides, to improve feature separability, several feature selection methods include GA and particle swarm-based optimization that have different drawbacks consisting of the tendency to fall into local optima and computational inefficiency [130]. The presence of redundant and insignificant features in the high-dimensional feature set can minimize fault identification accuracy. Moreover, unlike traditional classifiers, the artificial immune system (AIS) is a commonly used technique for anomaly detection, independent of the *a priori* knowledge of the model failure modes. For motor bearing faults, traditional fault detection methods are based on solving specific problems, and their design depends on various aspects of the bearing. Reference [129] proposed a GA optimization algorithm that combined unique multi-domain feature extraction with a supervised artificial intelligence technique (GA-AIS) to train the detector using the minimum amount of information. The proposed method can be used to detect fault types under a wide range of operating conditions. Generally, there is no universal deterministic approach that can be used for all practical fault situations. Negative selection algorithms (NSA) are used for detectors to distinguish standard data from faulty data. NSA and its variants analyze the vibration and current signals of the motor. Different detector generation strategies form NSA with different variants, including GA optimization, clone selection optimization, and stochastic native [131,132]. However, the increase of data leads to a decrease in the speed of detector generation and detection.

The main method of heredity in living things is replication. In the process of inheritance, the genetic material DNA of the parents is copied to the DNA of the parents, which is then copied to the offspring. This means that when a cell divides, the genetic material DNA is copied and transferred to the new cell which then inherits the genes from the old cell. When a sexually reproducing organism reproduces the next generation, two homologous chromosomes recombine by crossing over. This means that the DNA is cut at the same place on both chromosomes, and the two strands cross to form two new chromosomes. This means that the DNA of both chromosomes is cut at the same location, and the two strands cross to form two new chromosomes. During cell replication, it is possible, albeit very rarely, for some replication errors to occur, leading to some variation in DNA (such as DNA mutations). Some mutations can occur in the DNA, which creates new chromosomes. These new chromosomes exhibit new characteristics. GAs borrow many features of biological evolution, some of their main biological evolutionary features are that evolution occurs in the encoding of solutions. These codes are called chromosomes in biological terms. As a result of encoding, all properties of the optimization problem are studied through encoding. Encoding and decoding are the subjects of GAs. However, the law of selection determines which chromosomes produce more offspring than the average number, and in the GA, through the optimization problem, the fitness function is artificially constructed to achieve good chromosomes that produce more than the average number of offspring. Combining, the combination of genes from both parents enables the child to retain the characteristics of both parents. When chromosomes combine, random mutations cause offspring to differ from their parents.

The mysteries of genetics and evolution have not been fully solved. That is, their mechanisms remain not fully understood, and the intricacies of chromosomal encoding and decoding, as well as the methods of control, have yet to be fully comprehended. However, the following features of inheritance and evolution have reached a consensus: All the genetic information of an organism is contained in its chromosomes, which determine the characteristics of an organism. Chromosomes are composed of genes and their regular arrangements, and the process of inheritance and evolution occurs on chromosomes. The process of reproduction of an organism is carried out by the duplication of its genes. Through crossover between homologous chromosomes or chromosomal mutations, new species are created, and organisms take on new traits passed on to the next generation. Genes or chromosomes that are well adapted generally have a better chance of being passed on to the next generation than those that are less well adapted.

An optimization problem to find the maximum value of a function (and similarly to find the minimum value) can generally be described as the following mathematical.

The academic programming model:

$$\begin{cases} \max f(x) \\ \text{s.t.} \quad X \in R \\ \quad R \subseteq U \end{cases} \tag{3.64}$$

where $X = [x_1, x_2, \ldots, x_n]^T$ is the decision variable, and $f(x)$ is the objective function, Equations (3.1) and (3.3) are constraints, and U is the fundamental space, and R is a subset of U. A solution X satisfying the constraints is called a feasible solution. The set R represents a collection of all solutions satisfying the constraints, which is the set of feasible solutions. The relationship between them is shown in Figure 3.36.

As shown in Figure 3.36, for the above optimization problems, there are various objective functions and constraints, some are linear and some are nonlinear; some are continuous and some are discrete; some are unimodal and some are multimodal. In many complex situations, it is neither possible nor realistic to find the optimal solution exactly, so finding near-optimal or satisfactory factory solutions is one of the main focuses. Generally speaking, there are three main methods for finding the optimal or near-optimal solution: enumeration, heuristic, and search algorithms. As problem types vary and problem sizes increase, it remains a challenge to find a general approach that can solve these optimization problems at a limited cost. However, the GA provides an effective method and a general framework for solving such problems, creating a new type of global optimization search algorithm.

In genetic algorithms:

The n-dimensional decision vector $X = [x_1, x_2, \ldots, x_n]^T$ is depicted by a symbolic string X consisting of n notations $X_i(i = 1, 2, \ldots, n)$):

$$X = x_1 x_2 \ldots x_n \rightarrow X = [x_1, x_2, \ldots, x_n]^T \tag{3.65}$$

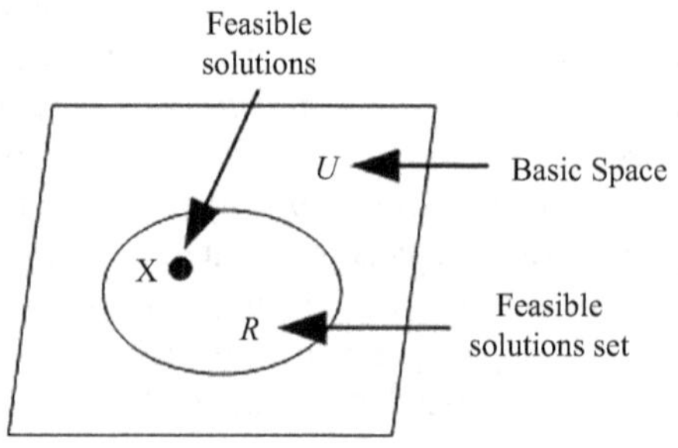

Figure 3.36 The diagram of planning model

By considering each x_i as a genetic gene, X can be thought of as a chromosome made up of n genes. Here the alleles can be a set of integers. It can also be a range of real values, or it can be purely a notation. The simplest alleles consist of two integers, 0 and 1, and the corresponding chromosome can be represented as a binary string of symbols.

The arrangement X resulting from this coding is the genotype of the individual, and the X value corresponding to it is the expression of the individual. For each X, its fitness is determined according to certain rules, and the fitness of an individual is related to its corresponding. The fitness of an individual is related to the value of the objective function of the individual's phenotype X. The closer X is to the optimal value of the objective function, the greater the fitness; conversely, the smaller the fitness. In a genetic algorithm, the decision variable X forms the solution space of the problem. The search for the optimal solution to the problem is performed by the search process for chromosome X. Thus, all chromosomes X form the search space of the problem.

3.9.1 The algorithm of genetic algorithms

3.9.1.1 Selection (replication):

Selection from the population $p(t)$ of the generation to the population $p(t+1)$ of the next generation according to the fitness of each individual, according to a $P(t)$, according to certain rules or methods. The individuals, in population $p(t)$, are randomly matched pairs, and for each pair, with a certain probability of exchange (called crossover probability) between parts of their chromosomes. For each individual in the population $p(t)$, individuals in the population change the value of a gene at one or some locus with some probability (called the mutation probability). The gene values on one or some motifs change to other gene values.

To better understand the calculation process of the genetic algorithm, the main execution steps of the simple manual simulation genetic algorithm are shown in Figure 3.37.

Example: Find the maximum value of the following binary function:

$$\begin{aligned} &\{ \max f(x_1,x_2) = x_1^2 + x_2^2, \text{s.t.} x_1 \in \{1,2,3,4,5,6,7\}, \\ &x_2 \in \{1,2,3,4,5,6,7\} \end{aligned} \tag{3.66}$$

3.9.2 Individual encoding

The genetic algorithm operates on symbolic strings representing individuals, so the variables x_1, x_2 must be encoded as a kind of symbolic string. In this problem, unsigned binary integers are used. Since x_1, x_2 are integers between 0 and 7, they are each represented by a 3-bit unsigned binary integer. The 6-bit unsigned binary number formed by joining them together forms the genotype of

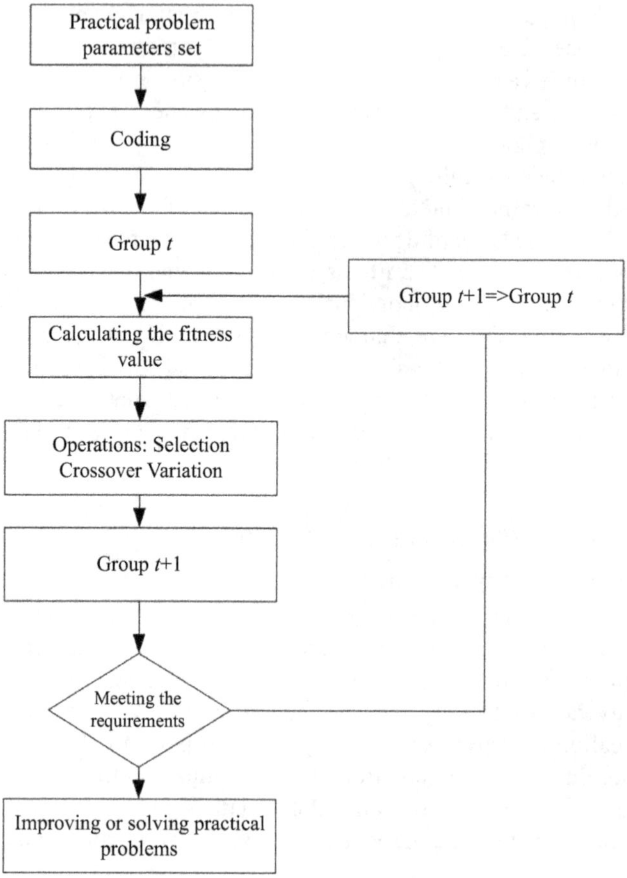

Figure 3.37 The operation process of genetic algorithm

the individual and represents a feasible solution. For example, the expression corresponding to the genotype: $X = 101110$ is $x = [5, 6]$. The expression type x and the genotype X of an individual can be interchanged employing coding and decoding procedures.

3.9.3 Generation of the initial population

The genetic algorithm is an evolutionary operation on a population, which needs to be given some initial population data to represent the starting search point. In this case, the population size is taken to be 4, that is, the population consists of four individuals, each of which can be generated by a random method:

For example: 011101, 101011, 011100, 111001

3.9.4 Fitness calculation

GAs use the magnitude of an individual's fitness to assess the strengths and weaknesses of each individual and thus determine its chances of inheritance. In this example, the objective function is always nonnegative, and the maximum value of the function is used as the optimization objective. In this case, the objective function is always nonnegative, and the maximum value of the function is used as the optimization objective.

3.9.5 Selection operations

As shown in Table 3.4, the selection operation (or replication operation) inherits individuals with high fitness from the current population to the next population according to some rule or model. It is usually required that individuals with higher fitness will have a greater chance of being inherited into the next generation of the population. In this example, we use a probability proportional to fitness to determine the number of individuals replicated in the next generation population.

This is done as follows:

1. Calculate the sum of the fitness of all individuals in the population at $\Sigma f_i(i = 1, 2, \ldots, M)$.
2. The relative fitness of each individual is calculated as $f_i/\Sigma f_i$, which is the probability that each individual will be inherited by the next generation population.
3. Each probability value forms a region where the sum of all probability values is 1.
4. A random number between 0 and 1 is generated, and individuals are selected based on which of the previous probability regions the random number appears in. The number of times an individual is selected is determined by which of the previous probability regions the random number appears in.

3.9.6 Crossover operations

As shown in Table 3.5, crossover is the main operation in genetic algorithms to generate new individuals, which exchange with each other with a certain probability between two individuals with a certain probability (Table 3.5).

Table 3.4 Selection operations

Individual no.	Initial group	x_1	x_2	Adaptability	Percentage	Selection times	Select results
1	011101	3	5	34	0.24	1	011101
2	101011	5	3	34	0.24	1	111001
3	011100	3	4	25	0.17	0	101011
4	111001	7	1	50	0.35	2	111001
Total				143	1		

Table 3.5 Crossover operations

Individual no.	Select results	Matching	Intersection points	Cross-sectional results
1	011101			011001
2	111001	1–2	1–2:2	111101
3	101011	3–4	3–4:4	101001
4	111001			111011

Table 3.6 Variation operations

Individual no.	Cross-sectional results	Variation points	Variation results	Offspring population $p(1)$
1	011001	4	011101	011101
2	111101	5	111111	111111
3	101001	2	111001	111001
4	111011	6	111010	111010

In this example, a single-point crossover is used, which is performed as follows:

1. Random pairing of the population.
2. Randomizing the location of the crossover points.
3. Some of the genes between the paired chromosomes are exchanged with each other.

3.9.7 Variation operations

Mutation is the process of changing the value of one or more genes in a locus of an individual with a small probability. It is also a method of generating new individuals.

In this example, we use the basic locus variation method to perform the mutation operation: First, determine the genetic variation position of each individual. Table 3.6 shows the position of the variation point randomly generated. The number in it indicates that the variation point is set at the locus. Then the original gene value of the variation point is reversed according to a certain probability.

A new generation of population $p(t+1)$ is obtained after a round of selection, crossover, and mutation operations on population $p(t)$.

As can be seen in Table 3.7, the population has evolved over one generation, with the maximum and mean values of adaptation. The table shows that after one generation of evolution, the population has improved significantly. The best individual "111111" has been found here. The genetic algorithm provides a general framework for solving complex system optimization problems, which does not depend on the specific domains that have strong robustness to the types of

Table 3.7 New generation of population

Individual no.	Offspring population $p(1)$	$x_1 \cdot x_2$	Adaptability	Percentage
1	011101	3 5	34	0.14
2	111111	7 7	98	0.42
3	111001	7 1	50	0.21
4	111010	7 2	53	0.23
Total			235	1

problems, so they are widely used in many disciplines. The following are genetic major application areas of the algorithm:

1. **Function optimization**
   Function optimization is a classic application field of genetic algorithms, and it is also a commonly used method for the performance evaluation of genetic algorithms. For some nonlinear, multi-model, multi-objective function optimization problems, it is difficult to find a better result that can be easily obtained by a genetic algorithm.
2. **Combinatorial optimization**
   As the scale of the problem increases, the search space of the combinatorial optimization problem also expands sharply, sometimes in the current calculation. It is difficult or even impossible to find the exact optimal solution by enumeration method on the computer. For such complex problems, people agree that realized that the main energy should be placed on seeking a satisfactory solution, and the genetic algorithm is the best tool to seek such a satisfactory solution. The practice has proved that genetic algorithm is very effective for NP-complete problems in combinatorial optimization. For example, genetic algorithms are already solving the traveling salesman problem, knapsack problem, bin packing problem, graph partition problem, and others have been successfully applied.
3. **Production scheduling problem**
   In many cases, the established mathematical model of the production scheduling problem is difficult to solve accurately. Even if it can be solved after some simplification, the solution result will be far from reality due to too much simplification. At present, in real production, scheduling is mainly based on some experience. Now genetic algorithm has become an effective tool to solve complex scheduling problems. GA has been effectively applied in single-piece production workshop scheduling, assembly line production workshop scheduling, production planning, and task allocation.
4. **Automatic control**
   In the field of automatic control, many problems related to optimization need to be solved, and the genetic algorithm has been initially applied to it and has shown good results. For example, use a genetic algorithm to optimize an aviation control system, use a genetic algorithm to design a space intersection

controller, optimize the design of a fuzzy controller based on a genetic algorithm, parameter identification based on a genetic algorithm, and learn of fuzzy control rules based on genetic algorithm. The structural optimization design and weight learning of the artificial neural network using the algorithm have shown the possibility of the application of the GA in these fields.

5. **Robotics**
   Robots are a kind of complex artificial systems that are difficult to accurately model, and the origin of genetic algorithms comes from the study of artificial adaptive systems, so robotics has naturally become an important application field of genetic algorithms. For example, genetic algorithms have been researched and applied in the path planning of mobile robots, trajectory planning of joint robots, solution of inverse kinematics of robots, structure optimization and behavior coordination of cellular robots, and others.
6. **Image processing**
   Image processing is an important research field in computer vision. In the process of image processing, such as scanning, feature extraction, image segmentation, and others, there will inevitably be some errors, which will affect the effect of image processing. How to minimize these errors is an important requirement to make computer vision practical. GA has found a place to use in the optimization calculation of these image processing and has been applied in pattern recognition, image restoration, image edge feature extraction, and so on.
7. **Artificial life**
   Artificial life is a man-made system that is simulated or constructed by artificial media such as computers and machinery and has the unique behavior of natural biological systems. Self-organization ability and self-learning ability are the two main characteristics of artificial life. There is a close relationship between artificial life and genetic algorithm, and the evolutionary model based on genetic algorithm is an important basic theory for studying the phenomenon of artificial life, although the study of artificial life is still in the enlightenment stage. However, the genetic algorithm has shown preliminary application ability in its evolution model, learning model, behavior model, self-organization model, and others, and it will be applied and developed more deeply. Artificial life and genetic algorithm complement each other. GA provides an effective tool for the research of artificial life, and the research of artificial life will also promote the further development of genetic algorithms.
8. **Genetic programming**
   Koza developed the concept of genetic programming. He used a coding method expressed in LISP language to automatically generate computer programs based on genetic operations on a tree structure. Although the theory of genetic programming is immature and its application has some limitations, it has been successfully applied in artificial intelligence, ML, and other fields.

9. **Machine learning**
   The ability to learn is one of the capabilities that an advanced adaptive system should possess. ML based on genetic algorithms, especially classifier systems, has been applied in many fields. For example, a genetic algorithm is used to learn fuzzy control rules, and a genetic algorithm is used to learn membership function, to better improve the performance of the fuzzy system; ML based on the genetic algorithm can be used to adjust the connection weight of ANN and can also be used. In the network structure optimization design of ANN, the classifier system has also been successfully applied in the learning multi-robot path planning system.

Chapter 4
Case study 1—AI-based electric motor fault diagnosis

The motor is the vital part of the electric vehicle (EV) powertrain system. With respect to the demand for motor performance in different environments, the motors should be endowed with fast speed, high output power, and high efficiency. Now, the mainstream motors used in EVs include PMSMs and induction motors (IMs) [137], which are mainly discussed in this section. It is important to ensure their correct operation.

As the working environments and working conditions of the EV motors are ever-changing, the motor parameters change constantly. For this situation, AI-based methods are well suited to accurately diagnose faults of different parts (stator winding, rotor PM, and bearing) of EV motors and can show marked performance [138]. The stator, rotor, and bearings will be described separately for the motor.

4.1 AI-based stator inner turn short-circuit fault diagnosis

Stator faults mainly include inner turn short-circuit (ITSC), phase-to-phase short-circuit, and ground short-circuit. According to surveys [139], the ITSC accounts for about 33% of all IM failures.

EV wheel traction motors have relatively high requirements for power/torque density, and their working conditions are complex and poor, which are more likely to cause serious ITSC fault. When ITSC is serious, it will evolve into phase-to-phase short-circuit fault and ground short-circuit fault, which will lead to high-temperature operation of the motor and irreversible demagnetization of the magnets, and catastrophic accidents will occur. Figure 4.1 illustrates the principle of ITSC fault, which assumes that one branch of phase C has ITSC faults and the fault winding in series.

For EV applications, the conventional method to detect ITSC faults is based on electromagnetic characteristics and thermal characteristics. Reference [140] introduces a Volterra kernel identification-based on-line fault diagnosis method, which refers to the variation of the stator branch voltage and current in the event of ITSC. However, the disadvantage of this type of method is that it is difficult to detect when ITSC failures are early or weak.

Reference [141] is a method based on the finite-element model and thermal network coupling model, which uses an open-loop physics-based back electromotive

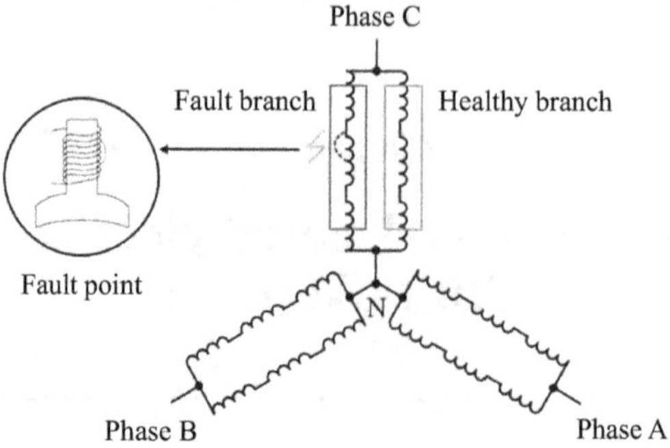

Figure 4.1 Schematic diagram of ITSC fault of PMSM stator

force (BEMF) estimator for EV PMSM ITSC fault detection. Similarly, in [142], the ITSC fault state of the PM alternator (PMA) was analyzed by the proposed electromagnetic-thermal-coupled model-based method. It can realize that the shortcoming includes lack of extremely strong prior knowledge and model knowledge, difficulties in weak fault identification, and limited fault degree identification.

Due to no prior knowledge of the process or its model parameters is required, these AI-based data-driven methods have recently been applied to resolve ITSC failures. It is analyzed here through a case study containing multiple representative methods. Reference [138] classifies ITSC faults based on the evidence of discrete wavelet energy ratio (DWER) by using Elman neural network (ENN). In [143], this researcher employs the dual-tree complex wavelet transform (DTCWT) to obtain fault features from vibration signals and import features into the probabilistic neural network (PNN) whose size is optimized by the orthogonal least squares regression (OLSR) algorithm. Reference [144] uses the Stockwell transform (ST) to extract different fault features in different frequency bands and proposes a parallel and different support vector machine (SVM) model to diagnose two ITSC faults simultaneously. The severity of ITSC can also be quantified. Reference [145] quantifies the level of ITSC failure for the line start PMSM (LSPMSM). It can be concluded that people are accustomed to supervised training on data through deep learning models to avoid model knowledge.

As obtained from Table 4.1, mentioned AI-based methods have a good ability to achieve high-precision early diagnosis of ITSC faults and multilevel fault diagnosis. Therefore, the motivation of employing AI-based methods is to make up for the shortcomings of traditional methods.

As for the disadvantage mentioned in [141], it is necessary to perform calculations and build a new EMF estimator for each machine before using the method in the actual EV motor. Through transfer learning and meta-learning technologies, the problem that the same estimator cannot take care of the diagnosis work of a large number of machines can be solved. This is a possible future trend for solving such faults.

Table 4.1 Performance of AI-based methods and suitable working conditions

Methods	Applied signal	Application targets	Working conditions	Top-one accuracy (%)
DTCWT-PNN [143]	Vibration	IM	Early faults	95.25
DWER-ENN [138]	Stator current	IM	Early faults and locating	≈1
SVM-based [144]	Stator current	IM	Two types of ITSC	96
CNN-based [145]	Stator current	LSPMSM	Four types of ITSC	97.75

In summary, for the ITSC fault, the training data of most AI-based methods are stator currents, the advantage of which is that it is very easy to measure and collect while EV driving.

4.2 AI-based rotor permanent magnet temperature monitoring

PM temperature is essential to ensure high performance and reliable control of PMSM for EV applications. Conventional PM temperature estimation methods can be divided into three major types: thermal model-based methods [146], BEMF-based methods [147], and high-frequency voltage injection-based methods [148].

Due to the rotation of the rotor, it is hard to measure the PM temperature, especially in actual EVs. In a laboratory environment, PM temperature can be measured by battery-powered equipment, infrared sensors, slip rings, and so on. These measurement methods cannot be performed in actual EV driving.

The data-driven PM temperature estimation method needs a huge amount of training data collected from the practical EV powertrains for the AI model [149]. Therefore, the application of AI in the PM temperature prediction of EV rotors still faces huge challenges.

The following part is a real project for estimating PM temperature by using generative adversarial network-supported random forest (GAN-RF) regression approach. The data used in this chapter comprise several sensor data collected from a PMSM deployed on a test bench. The PMSM represents a German OEM's prototype model. Test bench measurements were collected by the LEA department at the Paderborn University. The test bench parameters are shown in Table 4.2.

The data type of the dataset and their descriptions could be found in Table 4.3.

This dataset consists of multiple measurement sessions that can be between 1- and 6-h long and all recordings are sampled at 2 Hz. Boxplots help to present the distribution of quantitative data among features comparatively. The box in the figure shows the quartiles of the dataset, and the whiskers extend to show the rest of the distribution. The boxplots of the data are illustrated in Figure 4.2.

Pearson product-moment correlation coefficient (PPMCC) is a statistic proposed by the statistician Pearson to measure the degree of linear correlation between two random variables. As shown in (4.1), the PPMCC between two

Table 4.2 Introduction of test bench parameters

Parameters	Values
Power rating	22/52 kW
Voltage rating	177 V
Current rating	110/283 A
Torque rating	110/250 N m
Pole pair number	8
Cooling	Water–glycol
Inverter type	3×SKiiP 1242GB120-4DW
Typology	Voltage source inverter, two-level, IGBT
Interlocking time	3.3 μs
Controller hardware	dSPACE
Processor board	DS1006MC, four cores, 2.8 GHz

Table 4.3 Introduction of collected data types of the dataset

Ambient	Ambient temperature (in °C)
Coolant	Coolant temperature (in °C)
u_d	Voltage *d*-component measurement in *dq*-coordinates
u_q	Voltage *q*-component measurement in *dq*-coordinates (in V)
motor_speed	Motor speed (in rpm)
Torque	Motor torque (in N m)
i_d	Current *d*-component measurement in *dq*-coordinates
i_q	Current *q*-component measurement in *dq*-coordinates
pm	Permanent magnet temperature (in °C) measured with thermocouples and transmitted wirelessly via a thermography unit
stator_yoke	Stator yoke temperature (in °C) measured with thermocouples
stator_tooth	Stator tooth temperature (in °C) measured with thermocouples
stator_winding	Stator winding temperature (in °C) measured with thermocouples

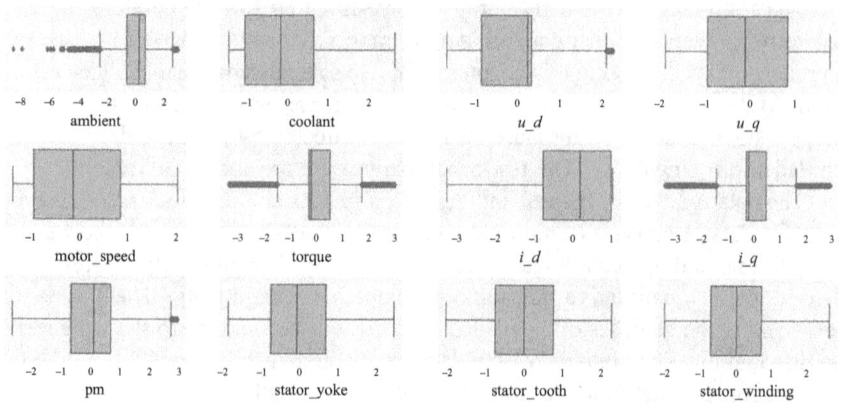

Figure 4.2 Boxplots of all features in the dataset

variables is defined as the covariance of the two variables $\sigma X_i, X_j$ divided by the product of their standard deviations $\sqrt{\sigma X_i, X_j \sigma X_j, X_i}$:

$$\rho X_i, X_j = \frac{\sigma X_i, X_j}{\sqrt{\sigma X_i, X_j \sigma X_j, X_i}} \tag{4.1}$$

Each element of the correlation matrix is a PPMCC. Different from the covariance matrix, each element of the correlation matrix is between $[-1, 1]$. Correlation matrix reflects the similarity of two variables per unit change. The correlation matrix of all features in the dataset is plotted by *seaborn* heatmap.

From the result of Figure 4.3, feature space should be limited to "ambient," "coolant," "*u_d*," "*u_q*," "motor_speed," "*i_d*," and "*i_q*." Conditional tabular generative adversarial network (CTGAN) is developed GAN model that can generate tabular data, which can learn the distribution of data better than Bayesian networks. This condition generator can overcome the problem of unbalanced training data and can help generate data with specific discrete values for data expansion. The CSV file is input into the CTGAN model as the input dataset. CTGAN model tabular data distribution and sample rows from the distribution. In this procedure, the mode-specific normalization is used to overcome the non-Gaussian and to process the imbalanced discrete columns. And then, the fully

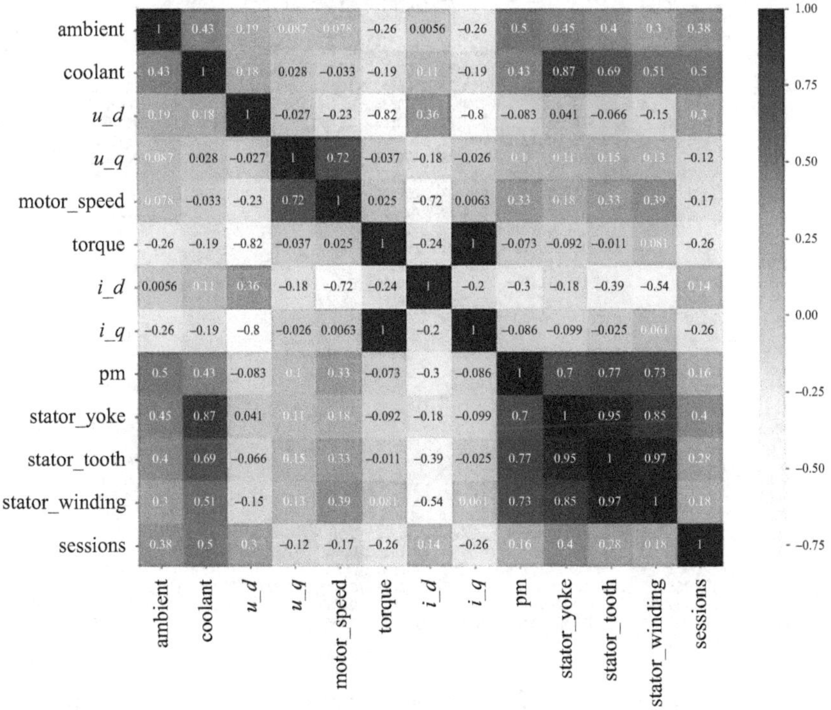

Figure 4.3 Correlation matrix of all features in the dataset

connected (FC) networks are employed to train a high-performance model. For the task in this chapter, the PMSM dataset should be inputted as the original data (shown in Figure 4.4(a)). The output of the CTGAN model is the generated new data (shown in Figure 4.4(b)).

This method employs RF algorithm to predict time-series data. Furthermore, data expansion by GAN improves model performance on PMs temperature prediction. The proposed method is shown in Figure 4.6. The result of the LSTM model (Figure 4.5) is a baseline for validating the results of the proposed methods.

The structure of the LSTM network used in this research employs seven network layers, including two LSTM layers, and five fully connected layers. In addition, two dropout functions are used in layers to avoid the risk of model training overfitting. Then, Figure 4.6 presents the suggested GAN–RF model.

Hyper-parameters are artificially adjusted parameters before or during training. A large number of hyper-parameters will have a direct and concise impact on the

ambient	coolant	u_d	u_q	motor_spe	torque	i_d	i_q	pm	stator_yoke	stator_tooth	stator_winding
-0.75214	-1.11845	0.327935	-1.29786	-1.22243	-0.25018	1.029572	-0.24586	-2.52207	-1.8314217	-2.0661428	-2.0180326
-0.77126	-1.11702	0.329665	-1.29769	-1.22243	-0.24913	1.029509	-0.24583	-2.52242	-1.8309687	-2.0648587	-2.0176313
-0.78289	-1.11668	0.332772	-1.30182	-1.22243	-0.24943	1.029448	-0.24582	-2.52267	-1.8304	-2.064073	-2.0173435
-0.78094	-1.11676	0.3337	-1.30185	-1.22243	-0.24864	1.032845	-0.24695	-2.52164	-1.8303328	-2.0631368	-2.0176322
-0.77404	-1.11678	0.335206	-1.30312	-1.22243	-0.2487	1.031807	-0.24661	-2.5219	-1.8304977	-2.0627947	-2.0181448
-0.76294	-1.11695	0.334901	-1.30302	-1.22243	-0.2482	1.031031	-0.24634	-2.5222	-1.8319309	-2.0625494	-2.017884
-0.74923	-1.11617	0.335014	-1.30208	-1.22243	-0.24791	1.030493	-0.24616	-2.52254	-1.8330117	-2.0621152	-2.0172427
-0.73845	-1.11399	0.336256	-1.30515	-1.22243	-0.24832	1.030107	-0.24603	-2.52284	-1.8321822	-2.0619526	-2.0172133
-0.73091	-1.11183	0.334905	-1.30379	-1.22243	-0.24778	1.029851	-0.24598	-2.52281	-1.8315759	-2.062443	-2.0177386
-0.72713	-1.10949	0.335988	-1.30563	-1.22243	-0.24829	1.029636	-0.24589	-2.52268	-1.8314383	-2.062317	-2.0181801
-0.72371	-1.10828	0.3354	-1.30456	-1.22243	-0.24791	1.029509	-0.24583	-2.52263	-1.8314928	-2.0620575	-2.0176919
-0.71775	-1.10859	0.334431	-1.30434	-1.22243	-0.24772	1.029386	-0.2458	-2.52264	-1.8318189	-2.0622487	-2.017435
-0.70488	-1.10999	0.336241	-1.30582	-1.22243	-0.248	1.029318	-0.24578	-2.52221	-1.8320519	-2.0621185	-2.0175893
-0.68253	-1.11136	0.335566	-1.30324	-1.22243	-0.24751	1.029274	-0.24578	-2.52193	-1.8320305	-2.061756	-2.0172567
-0.65458	-1.1126	0.336557	-1.30622	-1.22243	-0.24807	1.029222	-0.24575	-2.52214	-1.8324168	-2.0614564	-2.0170355
-0.63071	-1.11366	0.335352	-1.30556	-1.22243	-0.24759	1.029208	-0.24576	-2.52221	-1.8320504	-2.0615256	-2.017153
-0.6201	-1.11429	0.335909	-1.30568	-1.22243	-0.24788	1.029211	-0.24573	-2.52231	-1.8317783	-2.061331	-2.0176322
-0.61486	-1.1146	0.33629	-1.30662	-1.22243	-0.24791	1.02917	-0.24574	-2.52162	-1.8313637	-2.061172	-2.0182316
-0.60745	-1.11502	0.333016	-1.2344	-1.19087	-0.30744	1.0292	-0.24574	-2.52015	-1.8313625	-2.0611281	-2.0185127
-0.60374	-1.11589	0.324559	-0.98305	-1.07616	-0.34824	1.029236	-0.24573	-2.52024	-1.8324342	-2.0607681	-2.0184674
-0.60323	-1.11503	0.316031	-0.63691	-0.91523	-0.37626	1.029211	-0.24575	-2.52064	-1.8327438	-2.0605102	-2.0185668
-0.60046	-1.11193	0.308402	-0.26628	-0.7404	-0.39979	1.029245	-0.24574	-2.5211	-1.8335521	-2.0603254	-2.0184286
-0.59831	-1.10961	0.304178	0.091858	-0.56746	-0.37705	1.029249	-0.24574	-2.52254	-1.8346876	-2.0601928	-2.018966
-0.59977	-1.10908	0.299402	0.475354	-0.37738	-0.3843	1.029257	-0.24574	-2.52326	-1.8341513	-2.060098	-2.0199318
-0.60372	-1.10827	0.296587	0.874742	-0.17445	-0.40464	1.029674	-0.24575	-2.52304	-1.8329986	-2.0607262	-2.019915
-0.60357	-1.1074	0.296128	1.174401	-0.01128	-0.36782	1.011353	-0.24575	-2.52286	-1.8318759	-2.0616233	-2.0193489
-0.60256	-1.10541	0.290501	1.346797	0.105584	-0.33797	0.942681	-0.24574	-2.52299	-1.8307649	-2.062108	-2.0192761
-0.59803	-1.10299	0.278853	1.466556	0.21557	-0.37361	0.841679	-0.24574	-2.5229	-1.8299689	-2.0625234	-2.0192378
-0.59612	-1.10144	0.266134	1.549545	0.364731	-0.39639	0.658492	-0.24629	-2.52219	-1.8296397	-2.0620458	-2.0188093
-0.59272	-1.10003	0.261106	1.608851	0.528152	-0.365	0.457705	-0.24606	-2.52082	-1.8298637	-2.0618632	-2.0182965
-0.59473	-1.09925	0.258804	1.652615	0.6455	-0.33671	0.314666	-0.24597	-2.51996	-1.8298247	-2.0628257	-2.0183957
-0.59611	-1.10017	0.256707	1.683907	0.729556	-0.31627	0.212142	-0.24587	-2.52422	-1.8296678	-2.0632231	-2.0186977
-0.59491	-1.10153	0.254911	1.705966	0.789769	-0.30114	0.1391	-0.24581	-2.52375	-1.8296564	-2.0622692	-2.0171971
-0.60183	-1.10293	0.253605	1.722215	0.832909	-0.29081	0.086286	-0.24574	-2.52228	-1.8297886	-2.0625405	-2.016319
-0.61119	-1.10417	0.252927	1.733766	0.863816	-0.28345	0.048431	-0.24572	-2.52105	-1.830422	-2.0639482	-2.0179827
-0.62374	-1.10429	0.25259	1.741973	0.885961	-0.27825	0.0216	-0.2457	-2.52173	-1.8299776	-2.0640023	-2.0194201
-0.64057	-1.10407	0.252693	1.747813	0.901828	-0.27455	0.002424	-0.24569	-2.52371	-1.8291261	-2.0628276	-2.0199726
-0.65487	-1.1043	0.251911	1.751652	0.913199	-0.27129	-0.01076	-0.24566	-2.52394	-1.8287508	-2.0619857	-2.0191412
-0.65941	-1.10407	0.25156	1.754867	0.921344	-0.26949	-0.02046	-0.24562	-2.52381	-1.82847	-2.0613825	-2.0158405

(a)

Figure 4.4 Data used for training proposed method: (a) the first 40 rows of the input dataset of CTGAN, (b) the first 40 rows of the generated dataset by CTGAN model

ambient	coolant	u_d	u_q	motor_spe	torque	i_d	i_q	pm	stator_yok	stator_too	stator_win
-0.82786	-1.06194	0.18173	1.675528	2.024118	-0.26467	-0.81294	-0.25545	-1.1203	-0.94352	-0.76195	-0.73071
-0.87514	-1.09533	0.335988	1.675864	2.024117	-0.2653	-0.87591	-0.25557	-1.03623	-1.70488	-0.77636	-0.80711
-0.88789	-1.11695	0.184042	1.466556	2.024118	-0.26388	-0.804	-0.25581	-2.52064	-1.56836	-1.44471	-0.50321
-0.94053	-1.07996	0.180964	1.678925	2.024117	-0.26514	-0.81423	-0.25499	-0.88634	-0.97484	-0.76195	-0.53361
-0.93917	-1.09519	0.187872	1.679043	2.024078	-0.26774	-0.80517	-0.25548	-0.95887	-1.74629	-0.82244	-1.26879
-0.78094	-1.05139	0.183373	1.677201	0.941664	-0.26461	-0.7951	-0.25548	-0.95085	-1.82998	-0.76195	-2.01942
-0.99633	-1.04012	0.182984	1.677799	2.024118	-0.26723	-0.80623	-0.25615	-1.0094	-1.12709	-1.10211	-0.60685
-0.94522	-1.09492	0.185002	1.671634	2.024117	-0.26449	-0.82136	-0.25684	-1.64215	-1.04477	-1.16143	-0.50501
-0.59977	-1.06447	0.185557	1.675531	2.024119	-0.26501	-0.79245	-0.25523	-1.33739	-1.75501	-1.13972	-1.57884
-0.91217	-1.0656	0.186773	1.678528	2.024117	-0.26294	-0.87227	-0.25554	-2.27221	-1.78677	-0.68917	-0.72414
-0.77402	-1.08285	0.251339	1.679098	2.024118	-0.26612	-0.78839	-0.25597	-0.844	-1.28521	-0.95429	-1.68096
-0.79486	-1.10153	0.18439	1.677367	-1.22243	-0.26688	-0.79735	-0.25625	-2.51059	-1.13493	-0.76195	-2.01763
-0.85629	-1.06339	0.182132	1.678559	2.024117	-0.26547	-0.83473	-0.25639	-0.81973	-1.00384	-0.80243	-0.47634
-0.85433	-1.06949	0.183355	1.681903	2.024121	-0.26581	-0.04317	-0.25596	-1.01104	-1.83415	-0.76195	-1.86126
-0.96291	-1.05357	0.183812	1.68163	2.024118	-0.26533	-0.80942	-0.24564	-2.30487	-1.48105	-0.76195	-0.48596
-0.79835	-1.06221	0.181771	1.676728	2.024117	-0.37361	-0.81709	-0.25572	-2.00659	-1.8305	-0.76195	-0.47472
-0.93094	-1.05241	0.183384	1.674727	2.024118	-0.26747	-0.78524	-0.25561	-0.98646	-0.98934	-2.06205	-0.59237
-0.63423	-1.07634	0.183873	1.680168	2.024119	-0.26353	-0.83408	-0.25596	-1.68529	-1.24497	-0.77764	-0.59528
-0.92486	-1.10908	0.183975	1.678115	-1.22243	-0.26514	-0.79898	-0.25566	-1.32736	-1.03709	-1.38115	-0.96325
-0.92917	-1.05592	0.188443	1.671856	2.023688	-0.26476	-0.86438	-0.25536	-1.55628	-1.15234	-0.75985	-0.85898
-0.99337	-1.07455	0.187783	1.681116	2.024117	-0.26224	-0.81591	-0.24565	-1.82279	-1.03315	-0.76195	-0.977
-0.92784	-1.08383	0.181255	1.674836	1.935157	-0.2658	-0.84811	-0.25593	-1.68546	-1.35469	-0.83236	-1.57642
-0.93087	-1.04037	0.186342	1.677802	2.024118	-0.26743	-0.78637	-0.25542	-2.17215	-0.96161	-1.28297	-0.70656
-0.93012	-1.09207	0.180828	1.677802	2.024123	-0.365	-0.84759	-0.25599	-2.43629	-1.26763	-0.76214	-1.08675
-0.88185	-1.08429	0.184759	1.679601	2.024117	-0.26597	-0.84474	-0.25546	-0.83922	-0.95002	-1.06826	-0.47133
-0.91049	-1.05413	0.186964	1.681614	2.024118	-0.26574	-0.82605	-0.25641	-1.09298	-1.20074	-0.76195	-0.56539
-0.70327	-1.0575	0.18825	1.675674	2.024122	-0.26493	-0.82553	-0.25523	-2.5209	-0.99511	-1.0903	-1.34677
-0.99816	-1.10179	0.187833	1.677896	2.024118	-0.26597	-0.85893	-0.24561	-2.39021	-1.23377	-1.70983	-1.04935
-0.75569	-1.0824	0.189674	1.681229	2.024118	-0.26554	-0.81338	-0.25649	-2.03711	-1.06634	-0.82645	-1.24076
-0.99633	-1.05968	0.186884	1.672546	2.024118	-0.26619	-0.81483	-0.25567	-0.85793	-1.02006	-0.89744	-1.98521
-0.88228	-1.04984	0.185103	1.671634	2.024118	-0.26504	-0.81394	-0.25561	-2.06979	-0.97728	-0.76195	-1.50794
-0.92502	-1.07233	0.182882	1.68163	2.024078	-0.26703	-0.78654	-0.25571	-1.63202	-1.27275	-0.76159	-1.54166
-0.98832	-1.06927	0.181468	1.675281	2.024118	-0.26353	-0.86971	-0.2563	-1.2176	-1.74629	-1.15377	-0.53087
-0.94967	-1.0706	0.181202	1.674651	2.024119	-0.2647	-0.79735	-0.25576	-1.27004	-1.83203	-1.0679	-0.70064
-0.65941	-1.07755	0.182624	1.678595	2.02412	-0.26385	-0.80087	-0.25526	-1.0602	-1.04976	-0.76195	-0.99912
-0.94933	-1.07926	0.18291	1.680129	1.850847	-0.26541	0.086286	-0.25544	-1.71552	-0.94365	-1.8287	-0.77834
-1.01025	-1.11617	0.188787	1.68072	0.901828	-0.26516	-0.85794	-0.25557	-2.17671	-1.15444	-1.75366	-0.67059
-0.95074	-1.07233	0.188173	1.679397	2.024118	-0.26358	-0.82068	-0.25606	-1.04993	-1.02513	-2.02184	-0.59995
-0.95045	-1.04112	0.241533	1.678718	2.024118	-0.26497	-0.78503	-0.2456	-1.03882	-1.24497	-0.76195	-0.50476

(b)

Figure 4.4 (Continued)

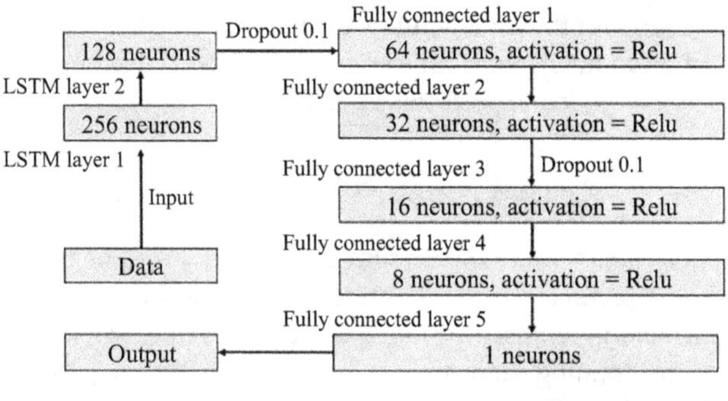

Figure 4.5 Structure of LSTM model used

performance of the model. Therefore, hyper-parameter optimization is a critical step for training machine learning models, the range of which includes learning rate, batch size, optimizer, as well as network width, number of layers, and connection method in deep learning. In this chapter, hyper-parameters of the RF model

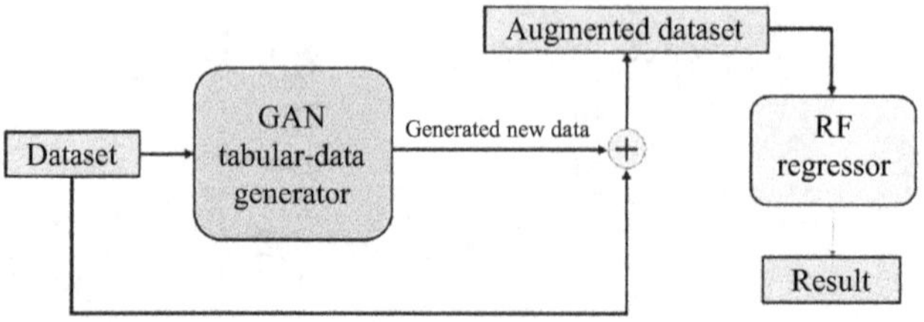

Figure 4.6 Diagram of proposed GAN-RF model

Table 4.4 Specifications of the computing device

CPU	Intel(R) Core(TM) i7-9750H CPU@2.60 GHz
RAM	16.0 GB
System Type	64-bit operating system
GPU	NVIDIA GeForce RTX 2060

and LSTM model are optimized by randomized search cross-validation (RandomizedSearchCV). After model optimization, the number of estimates is chosen as 54. The models in this chapter are trained on a computer. Specifications of the computing device are shown in Table 4.4.

The algorithms such as LSTM and RF used in the chapter are achieved in Python 3.7 with the support of TensorFlow–Keras and Scikit-learn framework. In the LSTM algorithm, the loss function is selected as the mean squared error (MSE), and the optimizer is the "Adam" optimizer that should be able to synthesize relatively strong. The batch size is 256, which is the stable algorithm performance, and the number of training epoch is 200. For the RF algorithm, the number of estimators is selected as the optimal value of 54 in the hyperparameter search process. Loss function still chooses MSE, the equation of which is presented as follows:

$$MSE(y_i, \widehat{y}_i) = \frac{1}{n}\sum_{i=1}^{n}(y_i - \widehat{y}_i)^2 \tag{4.2}$$

where MSE is the mean value of the sum of squared errors of the corresponding points between the predicted data and the original data. In (4.2), y_i represents the actual data value and $\widehat{y}_i$ represents the predicted data value, and n is the number of samples. After the training step, the temperature prediction performance of the GAN-RF method is tested with one test measurement session (the test curve is drawn as Figure 4.7).

Meanwhile, the performance of the GAN-RF method should be compared with that of the LSTM model (shown in Figure 4.8).

The MSE values on the test dataset of the two algorithms are shown in Table 4.5.

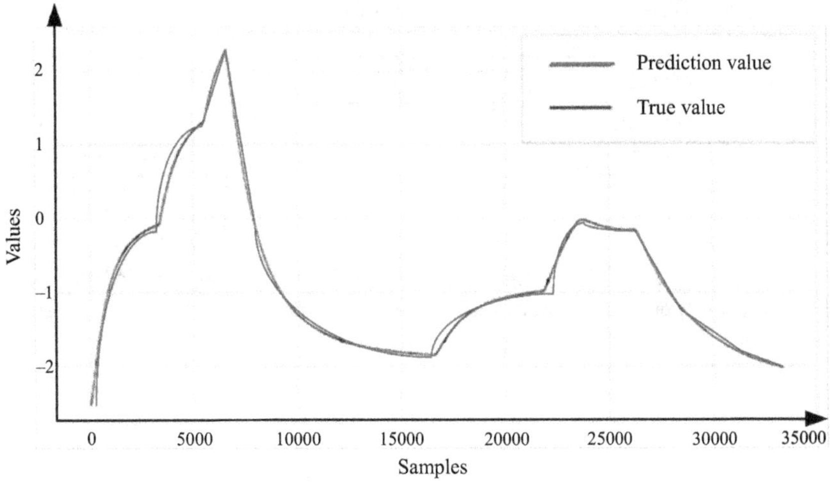

Figure 4.7 Test curve of GAN-RF model (the black line represents the true values of PM temperature; the red line represents the prediction values of PM temperature)

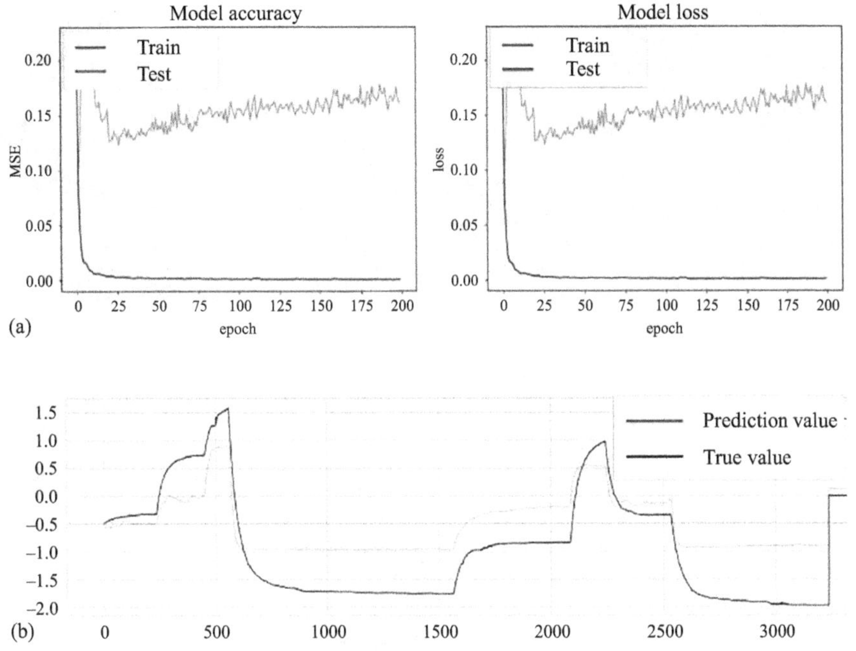

Figure 4.8 (a) Accuracy and loss of LSTM model; (b) test curve of LSTM model (the black line represents the true values of PM temperature; the red line represents the prediction values of PM temperature)

Table 4.5 Experiment results by MSE

Model	MSE
GAN-RF	0.0118
LSTM	0.3573

From the result of the comparison, it could be found that the GAN-supported RF algorithm gets the better performance on the task of PMs temperature prediction.

Chapter 5

Case study 2—AI-based inverter/IGBT fault diagnosis

The rapid growth of the EV industry makes people pay more attention to the safety of low-voltage and high-power inverters. In fact, the health of parts such as the power transistors in the inverter is closely related to the reliability of the entire device. The commonly used inverter fault diagnosis techniques mainly include three types: model-based methods, expert systems, and AI-based methods [150]. Therefore, the amount of literature on AI-based EV inverter methods is relatively small compared to other methods.

Currently, most of the power devices in EV inverters are insulated gate bipolar transistors (IGBTs). This is because the IGBT has the advantages of low state voltage drop and high input impedance. IGBT is taken as the main research object in this section. Figure 5.1 shows three periods of IGBT failure: early failure period, random failure period, and wear failure period with different failure rates.

AI-supported fault diagnosis methods in inverters for electric powertrains in EV involve the use of artificial intelligence techniques to detect and diagnose faults

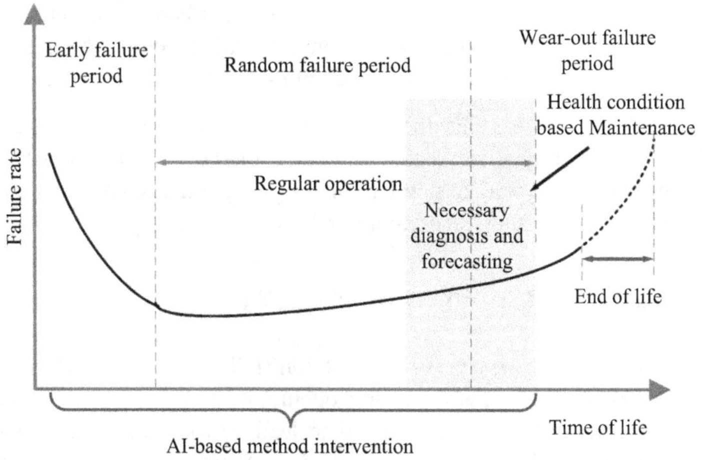

Figure 5.1 Power devices/IGBTs failure rate in lifetime and AI intervention timing

in the inverter system. Here are some common AI-supported fault diagnosis methods used in EV inverters:

1. Machine learning algorithms: Machine learning algorithms, such as support vector machine (SVM), random forests, or neural networks, can be trained on large datasets of normal and faulty inverter operation to classify and diagnose faults. These algorithms can learn patterns and identify deviations from normal behavior to detect faults in real time.
2. Signal-processing techniques: AI can be combined with signal processing techniques to analyze the electrical signals from the inverter system. Methods like fast Fourier transform (FFT) or wavelet analysis can be employed to extract relevant features from the signals, and AI algorithms can then be used to classify and diagnose faults based on these features.
3. Expert systems: Expert systems use knowledge-based reasoning to diagnose faults in the inverter. They rely on predefined rules and logic to interpret data and make informed decisions about the presence of faults. These rules can be developed by domain experts and encoded into the system.
4. Deep learning: Deep learning techniques, such as CNN or RNN, can be used to learn complex patterns in the data and perform fault diagnosis. Deep learning models excel at feature extraction and can automatically learn hierarchical representations from raw sensor data.
5. Data-driven approaches: AI can leverage large amounts of data collected from various sensors in the EV inverter system. This data can be used to build models that detect anomalies and diagnose faults based on deviations from normal operation. These approaches often involve unsupervised learning techniques such as clustering or anomaly detection.
6. Model-based approaches: AI can also be combined with physics-based models of the inverter system. By comparing the outputs of the physical model with the actual system behavior, deviations, and faults can be detected. AI algorithms can assist in real-time estimation and fault diagnosis by analyzing the discrepancies between the model predictions and actual sensor data.

It is important to note that the effectiveness of these methods relies on the availability of high-quality data, accurate models, and proper training. Additionally, domain expertise is crucial in designing and implementing these fault diagnosis methods to ensure accurate and reliable results.

5.1 AI-based RUL prediction for IGBT

EVs use IGBTs for efficient power conversion. IGBTs are exposed to electrical, thermal, and mechanical stresses during operation, and these stresses can cause performance degradation, which in turn can lead to wire-bond liftoff and solder fatigue. In EV applications, they may experience increased thermal cycling (temperature rise and fall), resulting in thermomechanical stress in the weak part of the IGBT package interconnection. Unexpected failures of IGBT will endanger the reliability of EV. By monitoring the trend of the fault precursor trajectory, this degradation can be identified at an early stage.

IGBT RUL estimation methods are roughly divided into physics-based analysis methods and AI data-driving methods. Among them, the physics-based method can provide a universal fault propagation method for the IGBT under test. However, these general physical models cannot represent the degradation of all IGBTs. Furthermore, due to the cumulative random behavior of thermal stress and mechanical stress, these methods have poor adaptability to dynamic environments.

The data-driven methods predicted by IGBT RUL are divided into conventional data-driven methods and AI-based data-driving methods. Both two types of data-driven methods can avoid complex physics and random uncertainty. Reference [151] proposes an RUL estimation approach based on the auxiliary particle filter (PF) to reduce the estimated variance by increasing the dimensionality of the sample and maintaining the diversity of the sample. Reference [152] also proposed a PF-based method for building a fault precursor by fusing the junction temperature (T_j) and collector–emitter on voltage ($V_{CE(on)}$). Reference [153] uses Gaussian process regression for applying a Bayesian inference on RUL estimation of the IGBT device.

Bayesian-based estimation methods can accurately track linear trends. PF can track nonlinear trends, but when the trajectory has both linear and nonlinear trends, their performance will decrease, especially when subjected to harsh measurement noise. In addition, these algorithms are computationally expensive and require a large number of samples to perform complex prior probability estimation. Reference [154] develops an algorithm that can be embedded and supports real-time decision-making on board. The probability of the state of health in the degradation process is obtained by performing a random process on the estimation result. Table 5.1 compares and analyzes the performance of the conventional data-driven methods and the AI-based method. Figure 5.2 presents the comparison of six feature values of the five measurable signals in EV inverter. In general, $V_{CE(on)}$ has a great comprehensive performance.

Table 5.1 Quantitative analysis of data-driven cases

Types	**Method**	**Experiment target**	**Reference parameter**	**Accuracy (RMSE)**
Conventional data-driven method	PF-based [152]	Simulation data from the IGBT module based on Hefner physical model from Saber software	Fusing $V_{CE(on)}$ and T_j	≤ 0.357
	APF-based [151]	Infineon Technologies $FS_{30}R_{0}6W_{1}E_{3}$ IGBT data	$V_{CE(on)}$	0.178
AI-based data-driven method	TDNN-based [154]	Power cycling tests data from Nottingham university	$V_{CE(on)}$	≤ 1.15

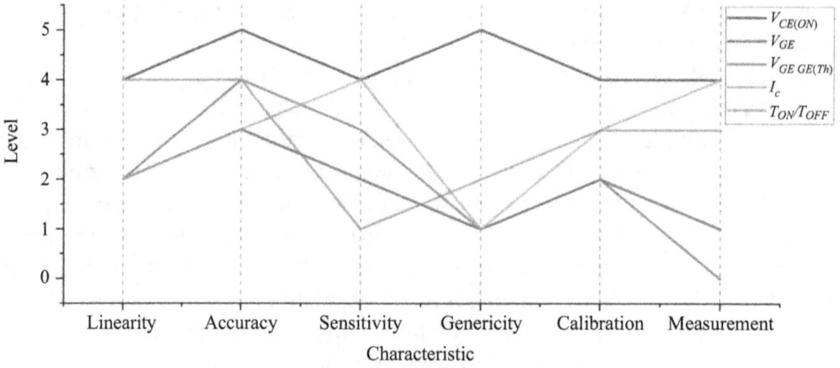

Figure 5.2 Performance comparisons of the fault evidence signals of IGBT ($V_{CE(on)}$ is the collector–emitter on voltage, V_{GE} is the gate–emitter voltage, $V_{GE(Th)}$ is the gate–emitter threshold voltage, I_c is the collector current, and T_{ON}/T_{OFF} donates the ratio of switch turn on time and turn off time).

The previous literature review summarizes and introduces some existing IGBT RUL prediction methods and characteristics of IGBT. The basic principles of each case will be analyzed later, with specific processes and specific code examples attached.

The aging of the module is related to many factors, including electrical load factors and environmental factors. Environmental factors include: temperature and humidity, cosmic rays, dust, mechanical vibration, and so on. Existing experiments and research are more on the influence of electrical load on IGBT life: The module is composed of different materials, and the CTE (coefficient of thermal expansion) of different materials is different. During the working process, continuous temperature fluctuations cause mechanical damage to adjacent layers of materials. Stress is the root cause of module aging failure.

First, we will introduce the two mechanisms of IGBT failure: sudden failure: that is, spontaneous and unpredictable failure; gradual failure: predictable failure, which occurs slowly over time, and the manufacturer plays a decisive role.

1. Sudden failure: The main task of the application engineer is to prevent sudden failure through design, and all relevant factors must be considered during design. It can be achieved through redundant systems and enhanced protection. Reasonable selection of devices and engineer's design determine the risk of sudden failure.
2. Gradual failure: The device manufacturer plays a decisive role. Generally, the manufacturer will provide device data charts, such as power cycle life and thermal cycle life, service life, and applicability.

The aging of IGBT devices can be modeled using various mathematical models. These models aim to capture the degradation and failure mechanisms that occur over time. Here are some commonly used traditional mathematical models for IGBT aging:

1. Arrhenius model: The Arrhenius model is based on the principle that the aging rate of a device is exponentially related to the temperature. It uses the Arrhenius equation, which relates the aging rate to the activation energy and temperature. The model assumes that the failure rate doubles for every 10 °C increase in temperature.

   Example Assuming an IGBT device has an activation energy (E_a) of 0.7 eV and the aging rate at 100 °C is known to be ten times higher than at 25 °C. Using the Arrhenius equation, we can estimate the aging rate at a different temperature.
2. Power law model: The power law model represents the relationship between aging and stress (such as voltage or current) applied to the IGBT device. This model assumes that the aging rate follows a power law function, where the aging rate is a power function of the stress level.

   Example Consider an IGBT module subjected to a constant voltage stress of 600 V. Based on experimental data, it is observed that the aging rate follows a power law relationship: Here, V is the applied stress voltage (600 V), V_0 is a reference voltage, and n is the power law exponent.
3. Coffin–Manson model: The Coffin–Manson model is commonly used to describe the fatigue behavior of materials subjected to cyclic stress. It relates the number of cycles to failure (N_f) with the applied stress level. This model assumes that the failure occurs due to accumulated damage caused by cyclic stress, and it predicts the device's fatigue life.

   Example Suppose an IGBT module is exposed to cyclic thermal cycling. Based on experimental data, the Coffin–Manson model can be used to estimate the fatigue life:

   Here, N_f is the number of cycles to failure, T is the range of temperature cycling, T_0 is a reference range, and b_{CM} is the Coffin–Manson exponent.
4. Weibull distribution model: The Weibull distribution is frequently employed to model the failure behavior of electronic components, including IGBTs. It provides a statistical description of the failure probabilities over time. The Weibull parameters, shape and scale, determine the failure characteristics, such as the initial failure rate and the time-dependent failure rate.

   Example Assume a set of IGBT modules undergo accelerated aging testing under specific stress conditions. By fitting the failure data to the Weibull distribution model, we can estimate the shape (beta) and scale (eta) parameters. These parameters can be used to calculate the probability of failure at different time points.

Figure 5.3 summarizes the different failure mechanisms of IGBTs. The gray box indicates the cause of sudden failure, and the shaded box indicates the factors that can lead to gradual failure. Failure may occur immediately or after a period of time (pre-damage), depending on the extent of the damage. In addition, cosmic rays are also an important factor for IGBT failure, which is a special failure mechanism and will not be reflected here.

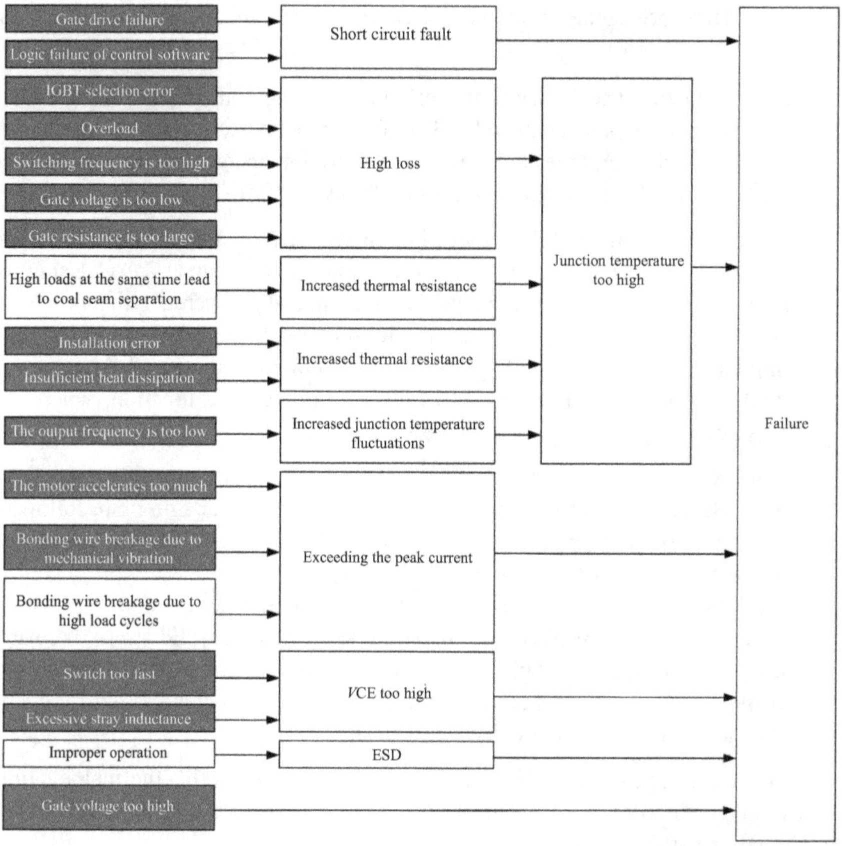

Figure 5.3 Failure reasons of IGBT

As shown in Figure 5.3, the failure reasons of IGBT can be summarized in detail as the following six points:

1. Excessive power dissipation: During operation, IGBT experiences conduction voltage drop and switching voltage rise, resulting in power dissipation. Prolonged exposure to high-power loads can lead to increased temperature, which can cause thermal failure.
2. Thermal failure: IGBT generates heat during operation. If effective heat dissipation is not provided, the temperature will rise continuously, exceeding the temperature limits of the IGBT and resulting in thermal failure. Thermal failure can lead to issues such as material expansion, poor contacts, and oxidation, affecting the device's conduction and switching characteristics.
3. Overcurrent damage: When IGBT is subjected to excessive current, it can be damaged due to overcurrent. Overcurrent can be caused by external circuit faults, short circuits, overloads, and so on. When the current exceeds the rated value of the IGBT, its internal structure may be damaged, leading to device failure.

4. Overvoltage damage: IGBT typically has a maximum voltage rating. If the device is subjected to voltage higher than its rated value, it can cause breakdown and voltage spikes, resulting in device failure.
5. Dynamic thermal failure: During the switching process, IGBT experiences switching losses and noise due to voltage and current variations. These factors can cause uneven temperature distribution in the IGBT, leading to thermal stress and expansion, resulting in failure.
6. Electrostatic discharge (ESD): ESD can cause breakdown and failure of IGBT. ESD typically occurs due to improper handling or inadequate ESD protection measures. Once an IGBT experiences ESD, its internal structure may be damaged, leading to device failure.

The failure and aging of IGBT modules are closely related. Over time and with changing operating conditions, IGBT modules may experience the following failure and aging issues:

1. Power module failure: The power chips within the IGBT module may fail due to overcurrent, overvoltage, or overtemperature conditions. These failures can be caused by external circuit issues, inappropriate operating conditions, or device faults. With increased usage time, power chips can undergo fatigue and degradation, leading to failure.
2. Solder fatigue and switching losses: The solder joints and connectors in IGBT modules can experience fatigue due to cyclic thermal expansion and contraction, resulting in poor contacts and failure. Additionally, the switching operation of the module can cause solder joint and connector losses, increasing the risk of failure.
3. Insulation aging: The insulation materials within IGBT modules can undergo aging over time. Factors, such as high temperature, humidity, and voltage stress, can cause degradation of insulation materials, leading to reduced insulation performance. Insulation aging can result in issues such as current leakage, breakdown, and short circuits, leading to module failure.
4. Thermal cycling stress: IGBT modules undergo temperature cycling during operation, causing thermal expansion and contraction of materials, leading to stress concentration and fatigue. Prolonged thermal cycling stress can damage the soldering, encapsulation materials, and internal structures, increasing the risk of failure.

Cycle curves are mainly power cycle curves and thermal cycle curves. One is concerned with junction temperature changes, and the other is concerned with case temperature changes. A small number of power devices do not mention the thermal cycle life because the thermal cycle life is much longer than the power cycle life.

1. How did the cycle curve come about? The cycle life is first designed. In the end, the research of power semiconductors is the research of physical characteristics, and then there is the research of production process. This part is mainly managed by the design department of power semiconductors. Second, the cycle curve is tested. Different temperature differences, different time

intervals, whether or not to add humidity, there will be various test conditions. Finally, some calculations of probability and statistics are added to obtain the circular curve given by the supplier. This part is mainly managed by the quality control department.

2. How to use the cycle curve? This actually tests the ability of the designer of the power electronic device to grasp the final application conditions. What kind of working conditions will appear and how often. What kind of working conditions will not appear etc. Some people say that my device is a general-purpose device, and the application conditions vary greatly, so I cannot grasp it. This question is also very simple. Everything is undecided, leaving a margin. This is why power electronics is a technology and not a science. Of course, there is another situation, the target use time of the equipment is very short, and it is a quasi-disposable machine. For example, the frequency converter used in the shield machine, such as the elevator in the building process. Just need to run for 2–3 years is enough. At this time, the focus is on the calculation and simulation of the junction temperature, and the life is relatively less important.

For providing a complete example of IGBT RUL prediction, we give an example by using Reference [155] and give the simple code.

5.2 AI-based IGBT open-circuit fault monitoring and detection

The open-circuit fault of IGBT is a relatively long-time problem, which will cause the whole system to gradually overheat and eventually collapse. This is also the reason why IGBT open-circuit fault is easier to detect and diagnose than short-circuit fault. When an IGBT is short-circuited, an abnormal short-circuit current usually occurs for only a few milliseconds, so it is difficult for the AI algorithm to detect a short-circuit fault. Even if the occurrence of a short-circuit fault is detected, it is difficult to take action to remedy before the system is damaged. Figure 5.4 illustrates the internal circuit principle of three cases of IGBT open-circuit fault, which presents that the current flows wrongly when open-circuit of T_1 happens in three cases.

At present, in the field of electrical inverter fault diagnosis, the mainstream diagnosis methods include the ones based on mathematical models of fault mechanisms, methods based on the expert system, and data-driven AI-based methods. The traditional methods have been applied and have achieved considerable results in diagnosing open-circuit faults. However, some shortcomings of these methods still exist. Model-based diagnosis methods need to establish a high-precision mathematical model describing the fault evolution mechanism. This method can achieve better results in actual applications. However, it is very difficult to establish a high-precision system fault model with complex multivariable. In addition, the established fault mathematical models are usually designed to solve specific device and are difficult to transplant to solve similar or other devices. Furthermore, with the increasing complexity of electromechanical devices, fault

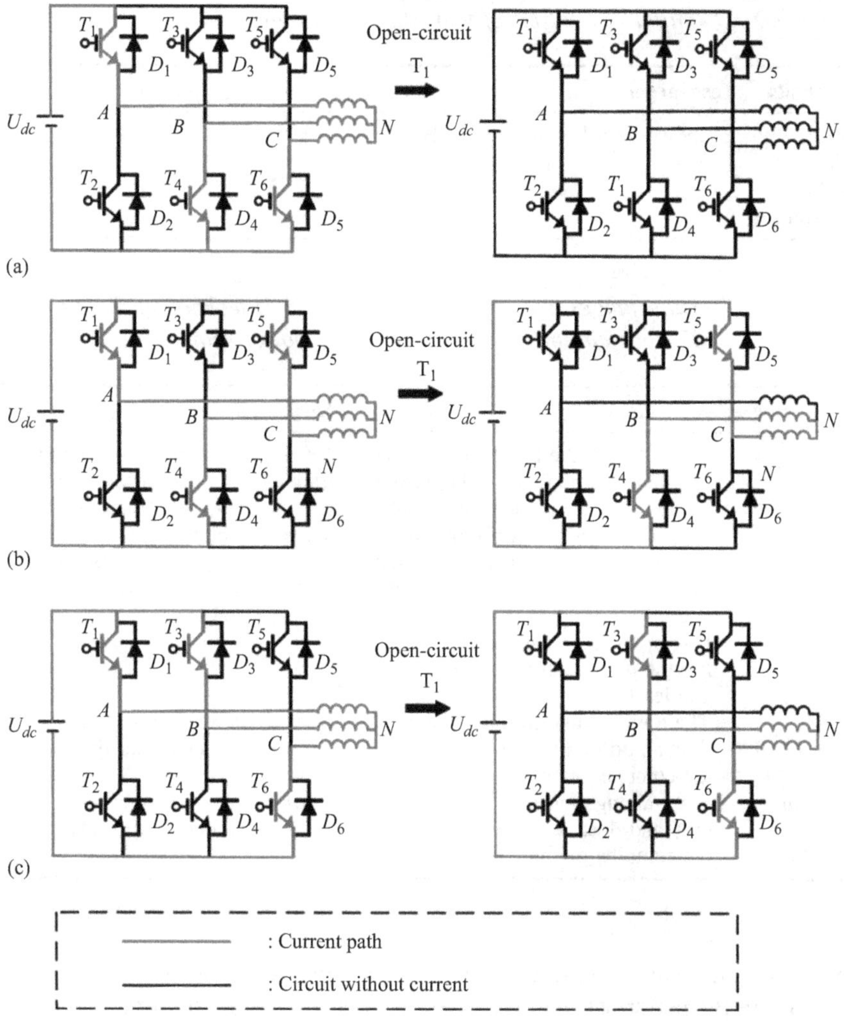

Figure 5.4 Schematic diagram of IGBT open-circuit fault principle (assuming T_1 open circuit)

diagnosis methods based on mathematical models are restricted in practical application. Therefore, with the rise of AI, people began to try intelligent diagnosis methods based on ML and NNs.

In the application of EV powertrain, SVM is a commonly used ML algorithm when diagnosing IGBT open-circuit faults. For example, Reference [156] uses SVM as a classifier to obtain fault types. Similarly, Reference [157] compares algorithms such as MLP, SVM, and *K*-means and believes that SVM can obtain better classification results. The difference is that [156] uses FFT to extract fault features from inverter output voltage, while [157] uses WT to extract fault

Table 5.2 Quantitative analysis of four ML technologies used in the case

Methods	Test target	Accuracy	Training time (s)	Complexity
MLP	EV powertrain VSI inverter	High	5	High
SVM		High	1	Low
SOM		–	5	Medium
K-means		–	2	Medium

Table 5.3 The advantages and disadvantages of four AI technologies used for IGBT open-circuit fault detection in EV powertrains

Technologies	Advantages	Disadvantages
MLP	Can solve all the problems with its complex and nonlinear relationships No limit on the number of outputs The high accuracy performance	Needs a very heavy and complex training. Due to split the data with a line, it has lowest margin of safety and highest risk
SVM	Low training process. Low programming complexity. Low implementation and testing time Due to the data separation, zone has the most safety marginalized and least risky	To solve very complex problems. Does not have a particular way to set the variables
SOM	The training algorithm is simple and easy and does not need to determine the complex output	Does not have accuracy in complex and nonlinear data
K-means	The training algorithm is simple and easy and does not need to determine the complex output	Does not have accuracy in complex and nonlinear data

information from three-phase currents of DC/AC inverter. Table 5.2 is a good case to compare the performance of AI-based methods (*K*-means, self-organizing map (SOM), MLP, and SVM) for EV powertrain voltage source inverter. The advantages and disadvantages of these ML technologies for EV inverter open-circuit fault detection can be summarized in Table 5.3.

Furthermore, for detecting open-circuit faults of neutral point clamped (NPC) inverters, Reference [158] uses modern compressed sensing theory to process the voltage signal into the proposed least-squares SVM with gradient information (G-LS-SVM) for fault classification. The author of Reference [159] innovatively uses the upper, middle, and down bridge voltages signals as the classification evidence for multilayer SVM. Differently, Reference [160] applies NN model that is used as the classifier.

The frequency of use of NN-based deep learning models in related research is gradually increasing [161–164]. Its most representative model is the ANN used in

Reference [165]. Other variants of NN have also received much attention. For example, the feedforward back propagation neural network (FFBPNN) in Reference [166] is used to diagnose IGBT open-circuit faults under variable load conditions at different frequencies and has good performance.

Diagnosis methods based on SVM, MLP, ANN, and other algorithms still require complicated and time-consuming manual extraction and marking of fault features, which could be avoided by deep learning algorithms with the ability to automatically extract features such as CNN. In [167], the author proposes to apply global average pooling (GAP) for modifying CNN model to diagnose the open-circuit fault of the DC–DC inverter, which extracts 12 types of fault features from signal-transformed images. We use the 12 types of IGBT open-circuit faults dataset to test the performance of various ML and DL algorithms, including SVM, *k*-NN, BPNN, DNN, CNN, and CNN–GAP (Tables 5.4 and 5.5).

Although the IGBT open-circuit failure will cause the entire system to collapse due to overheating, it is relatively easier to detect in terms of timing. IGBT in the inverter only withstands short-circuit current for a short period, even only a few milliseconds. Therefore, short-circuit faults are more difficult to detect through algorithms.

Here, an example Python code is given for providing a guidance to readers. It should be noticed that the key signal in this task is IGBT output current signals. In this example, we use sample data from normal operation and IGBT open-circuit fault. Each sample data are a time series of current signals, assuming there are 100 sampling points. The extract features function is used to extract features from current signals. In this example, we calculated statistical features such as mean, standard deviation, maximum, minimum, entropy, kurtosis, and skewness as example features. Then, we combine the extracted features with corresponding labels to form a complete dataset. After dividing the dataset into training and testing sets, we use an SVM as a classifier with a radial basis function kernel for training.

Table 5.4 Quantitative analysis of six AI technologies used in cases

Methods	**Test target**	**Average accuracy (%)**	**Training time (s)**	**Decision time (s)**
SVM	12 types of IGBT	96.13	22.79	4.14
KNN	open-circuit faults data	95.15	0.04	0.07
BPNN	from the DC–DC	80.32	78.19	0.05
DNN	inverter simulation	97.01	396.5	0.27
CNN	model	99.27	901.26	2.95
CNN–GAP		99.95	824.89	2.49

Table 5.5 Summary of contribution and AI algorithms of each reference with signals and feature methods

Fault type	Reference	Signal input	Feature method	AI algorithm	Contributions
Open-circuit	[158]	Voltage	–	G-LS-SVM	To use the compressed sensing theory to sparsely express the voltage fault signal
	[166]	Current	DWT	FFBPNN	• To be a new method of three-phase VSI fault diagnosis • Threshold needless under variable load conditions
	[160]	Voltage	JADE–ICA	NN	To use JADE-ICA algorithm to overcome the effects of nonlinearity and time difference
	[157]	Current	WT	MLP, SVM, SOM, K-means	To analyze five dynamic modes and five static modes
	[156]	Voltage	FFT, RPCA	SVM	To select the output voltages of the inverter as the fault characteristic signal
	[167]	1-D time-series voltage and current	CNN	CNN–GAP model	• To learn features directly from the original one-dimensional time-series data • To be fast and more accurate
	[161]	Three-phase current	ICA	NN	• To use ICA for three-phase current signal processing
	[159]	Upper, middle, and down bridge voltages	DFT	Multilayered SVM	To use a novel multi-layer SVM to detect the open-circuit fault of the inverter
	[162]	Upper, middle, and down bridge voltages	PCA	Multilayered ANN	To propose a novel multilayer neural network to diagnose all possible open-circuit faults
	[163]	Current	CNN	CNN	To learn fault features independently through a convolutional network
	[164]	Three-phase currents	Park's transform	BPNN	To apply Park's transform to obtain the three-phase current base wave amplitude as the feature variable for fault detection
	[165]	Three-stator currents	CEEMD	HHT-ANN	To use HHT to detect the harmonic characterizing the fault based on CEEMD of the three-stator currents

Chapter 6

Case study 3—AI-based bearing fault diagnosis

Mechanical components are an integral part of the electric vehicle (EV) powertrain system. The mechanical components mentioned, namely bearings and gearboxes, are indeed important mechanical components in EVs. They play crucial roles in the power transmission system. Here's the importance of these mechanical components in EVs:

Bearings: Bearings provide support and reduce friction in the electric motor and transmission system of an EV. Smooth operation of rotating components in the motor and transmission relies on high-quality bearings. Effective bearing design can reduce energy losses and noise while improving the system's lifespan and efficiency.

Gearboxes: Gearboxes are used in EVs to adjust the motor's torque output and speed to adapt to different driving conditions and requirements. Through proper gear transmission design, gearboxes provide high torque output and multiple gear options to meet power demands and deliver better driving performance.

The design and quality of these mechanical components are crucial for the performance and reliability of EVs. Their proper design and optimization can improve energy efficiency, reduce energy losses, and provide a better driving experience in EVs. Additionally, high-quality bearings and gearboxes can minimize friction and noise, extend the system's lifespan, and enhance the overall reliability of the vehicle.

The properties of these mechanical parts include durability, efficiency, precision, and so on. The durability of mechanical components refers to the need for high durability of mechanical components and the need to be able to withstand the high torque and high speed requirements of EVs. Moreover, they must be able to survive the EV's long-term operation without damage or excessive wear. Efficiency, on the other hand, refers to mechanical components that minimize energy loss. The high transmission efficiency of the gearbox in the process of transmitting torque means that its friction loss is low, which can ensure the effective conversion of electrical energy into mechanical energy. Mechanical components must be dimensioned with extreme precision. That's because precise mechanical structures are critical to the performance and stability of EVs, as well as reducing noise and vibration. For example, the gears in the gearbox need to be precisely matched and meshed to ensure the accuracy and reliability of torque transmission; the motor bearings need the precise size fit of the ball and the inner ring to minimize axial vibration.

The comprehensive consideration of these properties is crucial for the design and manufacture of high-performance, efficient, and reliable mechanical components for EVs. Manufacturers need to balance these characteristics when developing mechanical components that meet the demands of EVs and provide a great driving experience. Congenital manufacturing is important, but these mechanical parts will have some inevitable damage and failure during the working process, which will cause more or less property damage and even threaten human life. Therefore, real-time health monitoring and fault diagnosis of mechanical components in EV powertrain are crucial.

This chapter introduces two artificial intelligence (AI) applications in bearing system: bearing RUL prediction and multiple fault diagnosis.

6.1 AI-based bearing RUL prediction

RUL predictions of rolling bearings play a vital role in reducing unplanned maintenance fees and improving the reliability, availability, and safety of EVs.

In the RUL prediction of bearings, the following types of data are usually used: (1) vibration signal data: The most commonly used data type is vibration signal data for bearings. These signals can be acquired by accelerometers or vibration sensors. The vibration signal contains the running status and fault information of the bearing. (2) Temperature data: The temperature of the bearing is also an important indicator. Anomalies in temperature may indicate a fault or abnormal condition in the bearing. Therefore, temperature data can be used to predict the remaining life of the bearing. (3) Oil analysis data: Testing the chemical composition and physical properties of bearing oils can also provide information on the state of bearing health. Through oil analysis, abnormalities such as metal particles, moisture, acid value, and so on can be detected, which can be correlated with the remaining life of the bearing. (4) Working load data: Record the working load of the bearing, such as speed, load, and other parameters. Changes in workload can have an impact on bearing wear and life.

Combining the various types of data described earlier can provide a more accurate prediction of bearing remaining life. Typically, feature extraction and feature engineering are required on raw data to extract features related to bearing health status and remaining life. These features can then be used to train predictive models, such as regression models, deep learning models, or other machine learning models, to predict the remaining life of the bearing.

Bearing failure prediction methods mainly include conventional model-based, AI-based data-driven, and hybrid prognostics methods. Among them, the model-based approach is to establish mathematical and physical models to describe the degradation process of bearings. Recently, Reference [168] extracts features from the time-domain and time–frequency domain of bearing vibration signals and generates an extended Kalman filter estimator by learning and determining parameters. Qian *et al.* [169] proposed an enhanced PF and designed the RUL online prediction framework.

Table 6.1 Quantitative comparison of five different methods

Categories	**Methods**	**Test target**	**Top-one convergence speed rank**	**Top-one accuracy (CRA)**
Conventional model-based method	EKF-based [168]	Real experimental data of LDK UER204 rolling element bearings	2	0.88
	PF-based [169]		3	0.783
AI-based data-driven method	MODBNE [170]		5	0.901
	RVM-based		4	0.657
Hybrid prognostics method	[174]		1	0.961

However, the shortcomings and limitations of the model-based conventional method are that it requires *a priori* knowledge of the estimated parameters, and assumptions and simplifications will lead to uncertainty in the estimation. AI-based data-driven methods can solve this problem.

In [170], a multi-objective deep belief networks ensemble method is proposed, which uses a multi-objective evolutionary algorithm combined with traditional DBN training technology to simultaneously evolve multiple DBNs and combine the final evolved DBNs to build an integrated model for RUL estimation. In addition, RVM- and convolutional neural network (CNN)-based models are also used in some References [171–173] that can guarantee superior performance in the test.

Table 6.1 compares the performance of five state-of-the-art methods using the same vibration data from accelerated degradation tests of rolling element bearings. In the experiment, cumulative relative accuracy (CRA) is used to express the accuracy of the method. The closer the CRA value is to 1, the more accurate the RUL estimation result of the prediction method is. Analyzing the data in the case, the hybrid method seems to provide faster and more accurate RUL predictions.

It should be noticed that the use of AI-based data-driven methods for bearing RUL prediction is still a challenge because it is difficult to obtain marked life test data under real EV operating conditions. Almost all current AI-based research can only be verified in the laboratory.

Here also provide an example on how to train a bearing RUL prediction model by Gated recurrent units (GRUs). We first normalize the data and use *MinMaxScaler* to scale the data. Next, we convert the data into PyTorch tensors and define the hyper-parameters of the model, including the input dimension *input_dim*, the GRU hidden dimension *hidden_dim*, and the output dimension *output_dim*. Next, we created a RUL prediction model consisting of GRU and linear layers and defined the loss function and optimizer. After that, we created dataset and *DataLoader* for training and testing data. In the training loop, we use the training data for model training and calculate the training loss. After each epoch, we use the test data to calculate the test loss and output the current test loss. Finally, we use the trained model to make RUL predictions on new test samples. We first normalize new samples, then convert them to PyTorch tensors, and use the model to make predictions. Note that in practical applications, the code can be further modified and extended as needed, such as adding more hidden layers, tuning hyper-parameters, using different optimization algorithms, and so on, to achieve better RUL prediction performance.

6.2 AI-based multiple fault diagnosis

With the development of environment-friendly energy sources, EVs are gradually replacing fuel vehicles as the mainstream of the market. In PMSM drive systems for EVs, the rolling bearing is an essential mechanical component. In addition, rolling bearings are a fundamental component of rotating machinery and cause approximately 30% of failures in shaft drive equipment. Motor bearing failures can lead to serious consequences including motor response failure, abnormal noise, vibration, slow running, motor overheating, and even burnout. Failure to bearings can cause enormous hazards and losses and even threaten the safety of drivers. Therefore, real-time monitoring and diagnosis of their fault status enable appropriate repairs to be carried out in time to prevent further damage to the entire PMSM-drive system. Common factors in motor bearing failures include galvanic corrosion, inadequate lubrication, contamination, abnormal internal vibrations, and inadequate bearing loads.

In past research, bearing fault detection and diagnosis have been carried out by analyzing condition signals collected during bearing operation, including vibration, temperature, and acoustic emission. Vibration analysis is the main approach to bearing fault diagnosis due to the inherent characteristics of mechanical faults that can be contained in vibration signals.

The structure of the rolling bearing consists of inner race, outer race, the rolling element, and cage (Figure 6.1). The outer race is fixed to the bearing base and does not rotate with the rotating shaft. The inner race is connected to the drive shaft and turns together with the rotating shaft. Furthermore, in addition to the relative sliding of the rolling element against the inner and outer race, there is also a rotation of the rolling body. It is the key component for converting sliding friction into rolling friction. The cage is able to arrange the rolling elements at equal intervals to reduce friction and collisions between the rolling elements. Bearing vibrations are divided into two categories:

1. Periodic shocks caused by bearing faults.
2. Inherent vibrations of the bearing system.

When damage occurs to a rolling bearing, the bearing elements in contact with each other at the point of damage impact with it cyclically, resulting in low frequency vibrations. The fault frequency is related to the speed and geometry of the rolling bearing. When damage occurs to different parts of the bearing, a series of

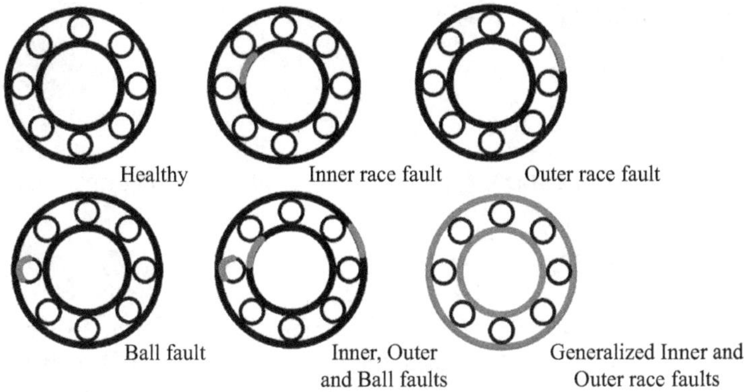

Figure 6.1 Regular fault types of ball bearing

cyclic vibrations at a specific frequency is generated, as well as the characteristic frequency of the failure. Frequency of rotation of bearings:

$$f_r = \frac{n}{60} \tag{6.1}$$

Rolling bearing inner race fault characteristic frequency:

$$f_{\text{inner}} = \frac{N}{2}\left(1 + \frac{d}{D}\cos(\alpha)\right)f_r \tag{6.2}$$

Rolling bearing outer race fault characteristic frequency:

$$f_{\text{outer}} = \frac{N}{2}\left(1 - \frac{d}{D}\cos(\alpha)\right)f_r \tag{6.3}$$

Rolling bearing rolling element fault characteristic frequency:

$$f_{\text{ball}} = \frac{D}{2d}\left(1 - \left(\frac{d}{D}\right)^2 \cos^2(\alpha)\right)f_r \tag{6.4}$$

The previous formulae are the characteristic frequencies of faults when faults occur in the individual components of rolling bearings. Deviations between actual and theoretical values occur, so in practical fault diagnosis, it is necessary to find an approximation to the characteristic frequency in the vicinity of the theory. When the bearing is operating, regardless of whether a fault occurs, the vibration generated is called the inherent vibration of the bearing system. Inherent vibration occurs due to the elasticity of the system, which mainly refers to the vibration of the bearing components inherent frequency decay. In addition, the inherent frequency is also related to the material, geometry, and dimensions of the rolling bearing.

Rolling bearing inherent frequency:

$$f_{\text{ball}} = \frac{0.24}{r}\sqrt{\frac{E}{2\rho}} \tag{6.5}$$

Intrinsic frequency of the inner and outer race of the bearing (material is steel):

$$f_{i,o} = 9.4 \times 10^5 \times \frac{h}{D^2} \times \frac{n(n^2 - 1)}{\sqrt{n^2} + 1} \tag{6.6}$$

The intrinsic frequency of a bearing element is likely to be in the range of 20–60 kHz, which is usually used as a diagnostic frequency band. This is because an impulse shock caused by a defective bearing element or its irregular surface during operation can excite a component of the bearing to vibrate at its intrinsic frequency, thus amplifying the fault signal in this band and increasing the signal-to-noise ratio of the fault signal.

Due to poor lubrication and mixed with foreign matter and other reasons, hence the rolling element surface deterioration, resulting in the original rolling surface unevenness become more powerful, this unevenness still has the randomness. Therefore, the vibration caused by this also maintains its randomness. The concave and convex shapes become larger, the corresponding excitation force will also increase, and the resulting vibration amplitude also increases accordingly, as shown in Figure 6.2.

Motor bearing damages are usually divided into six main types: fatigue, wear, corrosion, electrical erosion, plastic deformation, and fracture and cracking [175]. Each type of damages can have different development states and lead to sudden

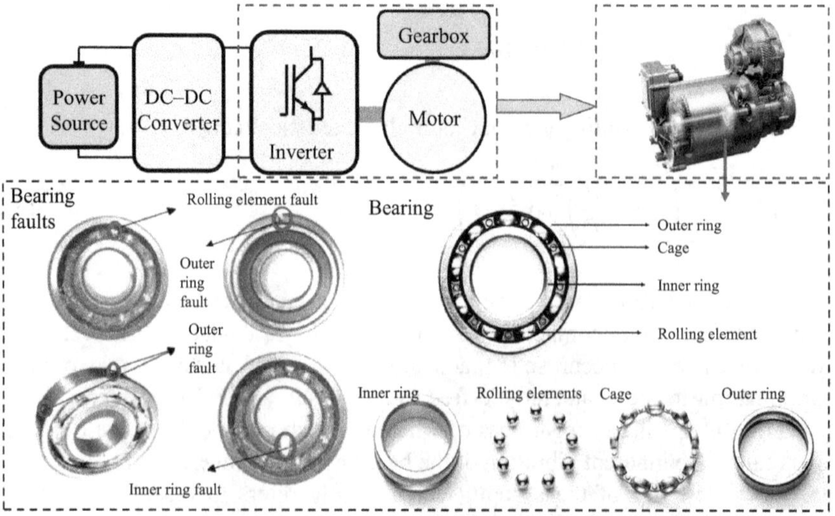

Figure 6.2 Rolling bearing faults in the structure of EV powertrains

failure [176]. Rolling element bearing is one of the most used bearings in EV motors. Figure 6.2 provides the example of bearing failures and faults positions on bearing. As shown in Figure 6.2, bearing damages can occur at the inner ring, rolling elements, cage, and outer ring.

Recent conventional methods for bearing faults diagnosis mainly include dictionary construction-based methods [177,178] and the sparse denoising methods in time-domain and time–frequency domain [179]. These methods have unavoidable disadvantages in EV applications:

1. The calculation process of the dictionary construction-based method is complicated and time-consuming, and it is difficult to diagnose online real-time faults in EVs.
2. Building a sparse denoising model for bearing fault diagnosis requires prior knowledge of bearing parameters, which may hinder its application in the EV industry.
3. These conventional methods have many hyper-parameters, which makes it difficult to maintain the models.

Therefore, relying solely on traditional PE methods, it is difficult or impossible to clearly detect the specific fault location and fault type. However, during EV driving, it is necessary to know the health status of the bearing in real time to avoid driving accidents in time.

The commonly used machine learning classifiers suited for EV bearing issues are support vector machine (SVM) and RF. They are usually only used to classify faults based on existing features but cannot directly extract deep fault features from the signal. This existing feature needs to be manually extracted from the initial signal using some signal processing methods and then input into the classifier for classification work [180–183]. However, AI-based approaches that combine feature extraction with AI classification algorithms not only consume a lot of time and manpower for manual data labeling before model training but also lack flexibility for different machines.

CNN is a neural network that has attracted much attention due to its powerful functions. Convolutional and pooling layers in CNN have powerful automatic feature extraction capabilities, while fully connected layers are generally applied for fault identification. The number of layers depends on the complexity of the application task. Here are two typical cases.

Reference [184] uses a combined method of CNNs and machine learning classifiers. On the one hand, LeNet-5 based convolutional layers extracts the shallow to deep fault features of time-domain vibration signals; on the other hand, multiple independent RF classifiers are separated by multiple local and global feature information obtained in different convolutional layers. Figure 6.3 shows the flowchart of the proposed signal-to-image method in Reference [184]. In [185], a special variant of CNN, Pythagorean spatial pyramid pooling CNN, can also deeply extract and analyze the deep fault information from continuous wavelet transform (CWT) scalogram, so as to achieve high-precision fault diagnosis under the condition of variable speed and full speed.

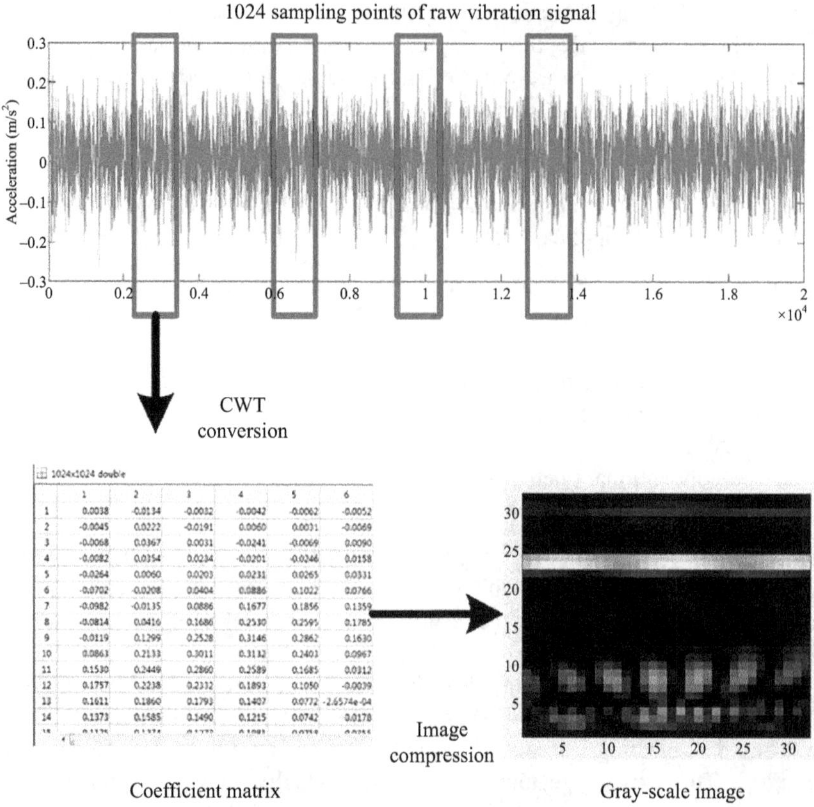

Figure 6.3 One example of method for signal-to-image conversion [184]

Model-based CNN could classify fault types directly [185,186]. Meanwhile, CNN could also be a part of diagnosis model as well. In [187,188], CNN is just as a powerful feature extractor. Therefore, SVM and NNs can also be followed by the convolutional layers for classification of fault types [184,187]. Figure 6.4 illustrates fault diagnosis process of convolutional layer with ML classifier. The ML classifier is used to accurately classify the deep fault feature information extracted and output from the feature image by the 2D CNN model. In [188], the author combines deep neural network (DNN) and convolutional layers to get a fault diagnosis architecture convolutional neural network with ensemble deep neural network (CNNEPDNN) for bearings, which recognize faults according to features with different discriminative powers. To address the gradient decay problem of deep CNN models, References [189,190] combine residual structures with convolutional layers to avoid gradient disappearance during CNN training.

In order to show the difference among various signal-to-image conversion methods, this chapter uses the bearing vibration signal in the Case Western Reserve University (CWRU) dataset to demonstrate the difference between methods and is summarized in Figure 6.5. These methods also work in other collected signals.

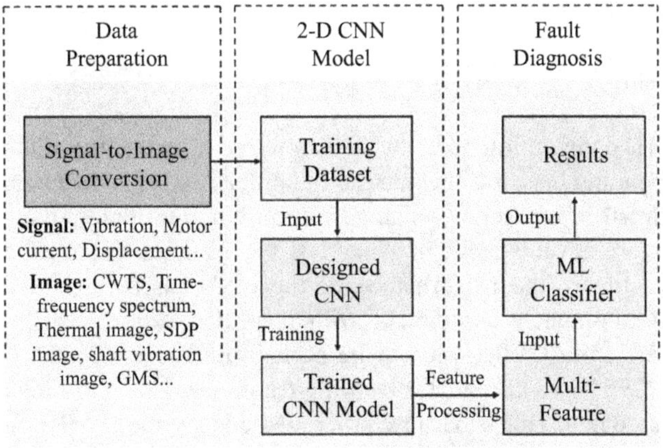

Figure 6.4 Flowchart of 2D CNN and ML classifier-based fault diagnosis structure

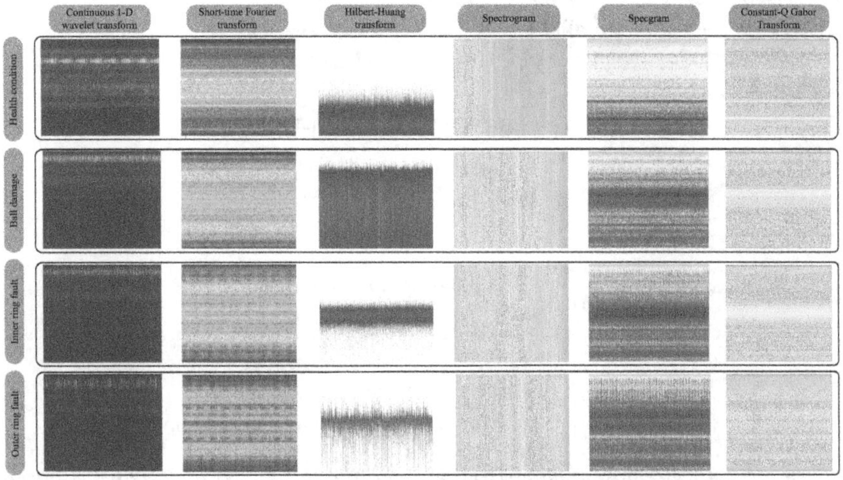

Figure 6.5 Bearing vibration signal conversion using different signal-to-image methods (CWRU rolling element bearing vibration signal)

The 1D CNN model was used to extract the deep features and final classification of the one-dimensional signals in References [191,192]. 1D CNN can directly process 1D bearing vibration signals, eliminating the need for a complex signal imaging step. Thence, compared with 2D CNN, 1D CNN-based model is more suitable for EV real-time applications. The 1D convolution is to take a dot product between a kernel $w_c \in R^m$ and the jth segmented signal $s^i_{j-m+1:j} \in R^m$ to get convolution features:

$$c_j = \text{Relu}\left(\sum_{i=1}^{n} w_c * s^{i}_{j-m+1:j} + b_c\right) \quad (6.7)$$

where * is a 1D convolution operator, w_c is referred to as the convolution kernel, b_c is the corresponding bias, n is the number of kernels, and c_j is the jth output point of the convolutional layer. Relu$(\cdot)$ is an activation function. Figure 6.6 explains why 1D CNN is more advantageous in actual EV applications. Table 6.2 aims to summarize and compare the performance of the CNN-based method of the above method in the computer experiment with the same dataset which is a labeled bearing dataset from CWRU. It can be seen from Table 6.2 that 1D CNN, the accuracy of 1D CNN with transfer learning technique, is slightly lower than other CNN-based methods. This is because other methods use the CWRU dataset as both training and verification data, but the diagnosis model in Reference [191] is a general model which transfers fault-related knowledge from other bearing systems and tests directly on the CWRU dataset.

Data are a vital part of AI-based diagnosis methods. Due to the difficulty of collecting the bearing fault data for a specific type of EV, the most suitable AI technology would be transfer learning. Transfer learning makes the model trained in the source domain still applicable in the target domain and greatly saves the time

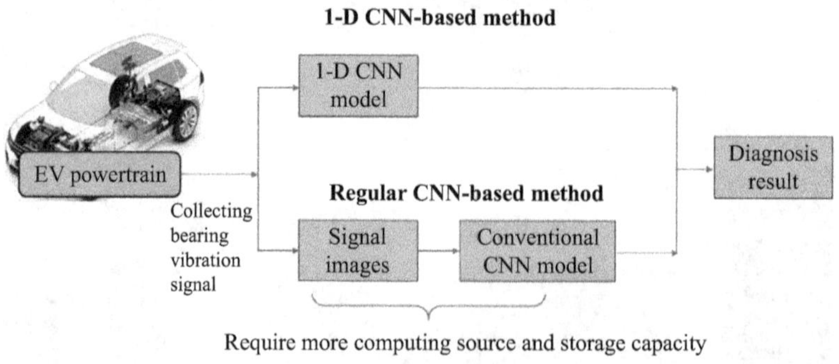

Figure 6.6 Architecture comparison between regular CNN- and 1D CNN-based method

Table 6.2 Performance of CNN-based methods under CWRU bearing data

References	Methods	Test target	Top-one accuracy (%)
[184]	CNN + RF	CWRU dataset	99.08
[185]	PSPP–CNN		97.79
[188]	CNNEPDNN		98.1
[187]	CNN + SVM		98.75
[191]	1D CNN + transfer learning		90.8

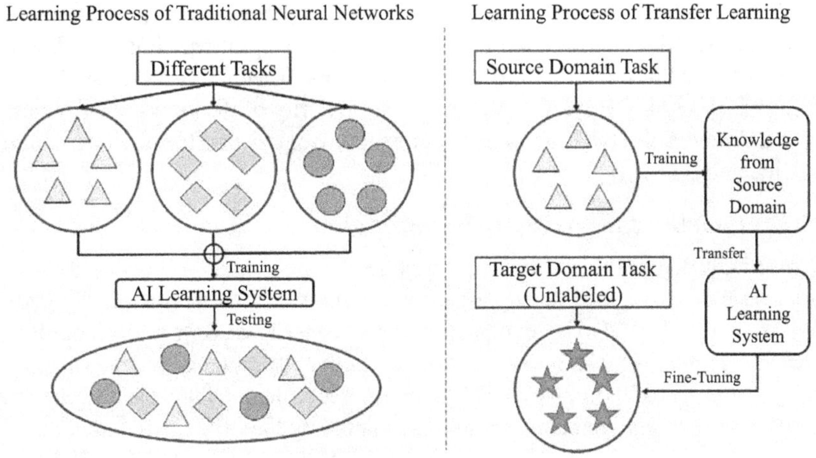

Figure 6.7 The principle of transfer learning and the difference with traditional NNs

Table 6.3 Performance of transfer learning-based method based on CWRU dataset

References	Methods name	Feature learning	Top-one accuracy (%)
[191]	DCTCL	Directly from vibration signal	90.8
[193]	DDC	Directly from vibration signal	88.5
[194]	DANN	Directly from vibration signal	88.5
[195]	TCA	Manually extract features	42
[196]	DAFD	Vibration signal spectrum	54.9

to construct a bearing health recognition model. Figure 6.7 demonstrates the principle of transfer learning.

For transferring fault diagnosis tasks where there are no available labeled data in the target domain, the method based on transfer learning is better than the classic method without transfer learning. Reference [191] proposed a deep convolutional transfer learning network (DCTLN) based on transfer learning technology, which can learn features directly from vibration signals.

AI-based methods should preferably use the same dataset for comparative analysis. Therefore, Table 6.3 analyzes and compares five transfer learning methods suitable for EV bearing fault diagnosis on the CWRU bearing dataset, which includes deep domain confusion (DDC) [193], domain adversarial training of NNs (DANN) [194], transfer component analysis [195], and deep domain adaptation neural-network-based fault diagnosis (DAFD) [196]. It can be concluded that transfer learning is a promising tool in the field of EV bearing diagnosis. In addition, the DCTCL method has relatively high accuracy and is suitable for intelligent fault diagnosis of unmarked data of EV bearings.

Thermal images-based AI diagnosis is an advanced and novel attempt for motor bearing failure. For example, in Reference [197], thermal images of bearing failure can be analyzed directly by the proposed modified CNN model without any signal preprocessing. However, due to the particularity of the working environment of the EV motor and the high integration of EV components, this method is hardly implemented on the EV currently.

6.2.1 Bearing fault diagnosis: example 1

Bearing fault diagnosis is a relatively mature research field. A bearing multi-fault diagnosis framework will be explained in detail here. The inferable deep distilled attention network (IDDAN) diagnosis method consists of four main modules (as shown in Figure 6.8): data augmentation (DA), backbone network, distillation strategy, and transfer inference. Transfer inference helps IDDAN become more adaptable to the target domain through fine-tuning by the very small amount of data after the pretraining stage. Particularly, the first and second parts of Figure 6.8 show how the bearing vibration signal is collected and preprocessed in the context of the current application.

The DA method of IDDAN's pretraining data in the source domain adopts multi-scale and multi-timescale signal conversion which is a special DA method proposed for the collected 1D bearing vibration signals.

Multi-scale signal conversion is similar to the idea of cropping and zoomed-in computer vision, which can improve the generalization ability of pretrained IDDAN. In other words, the domain adaptation ability of the IDDAN can be improved when transferring knowledge from the source domain to the target domain. Defining the length of each vibration signal as L_s and the sampling rate of as f_s, the principle of data enhancement can be expressed as Figure 6.9.

It can be obtained from Figure 6.9 that the strategy of DA could be presented by the following equation:

$$\left\{k \times \frac{L_s}{k}, k \in N^* \quad V_{\text{new}}(t) = \sum\nolimits_{n=-\infty}^{\infty} V_s(nT)\delta(t - nT) \right. \tag{6.8}$$

where k is the number of segments after each signal cutting, $V_{\text{new}}(t)$ is down sampled vibration signal, $V_s(t)$ is the original collected vibration signal, and δ represents the impulse function. In this DA step, collected 1D vibration signals are mainly processed in two ways:

1. Each vibration signal used for IDDAN pretraining step is segmented multiple parts, and the size of each segment is $\frac{L_s}{2}, \frac{L_s}{3}, \frac{L_s}{4}, \frac{L_s}{5}$, etc.
2. Each vibration signal used for IDDAN pretraining step is resampled by sampling rate $\frac{f_s}{2}, \frac{f_s}{3}, \frac{f_s}{4}$, etc.

Through this DA method, the amount of data can be increased efficiently, providing sufficient data for IDDAN pretraining to improve the effect of knowledge transfer.

The bearing health status recognition is realized through self-attention (SA) based network module as the main body. The backbone of IDDAN is a modified

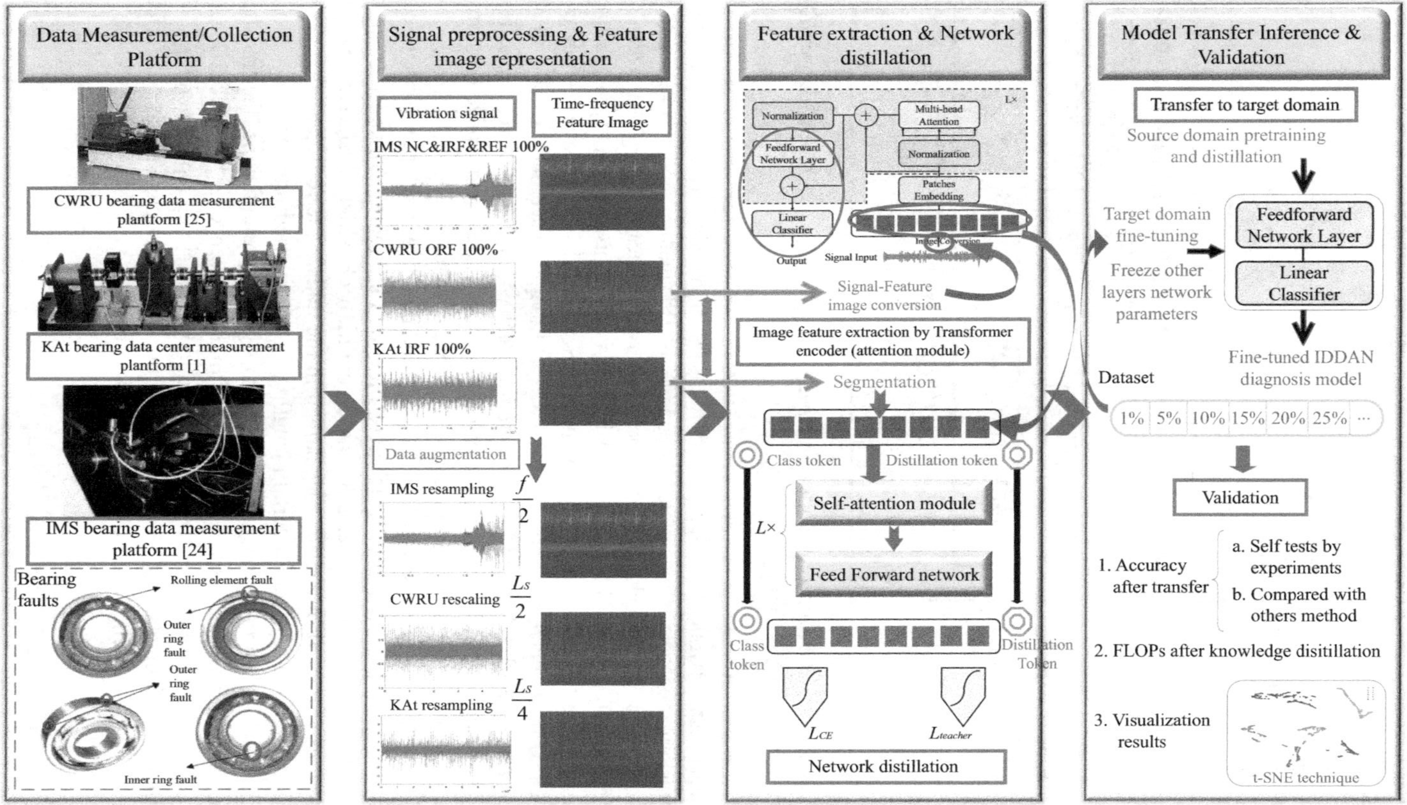

Figure 6.8 The architecture demonstration of the proposed method

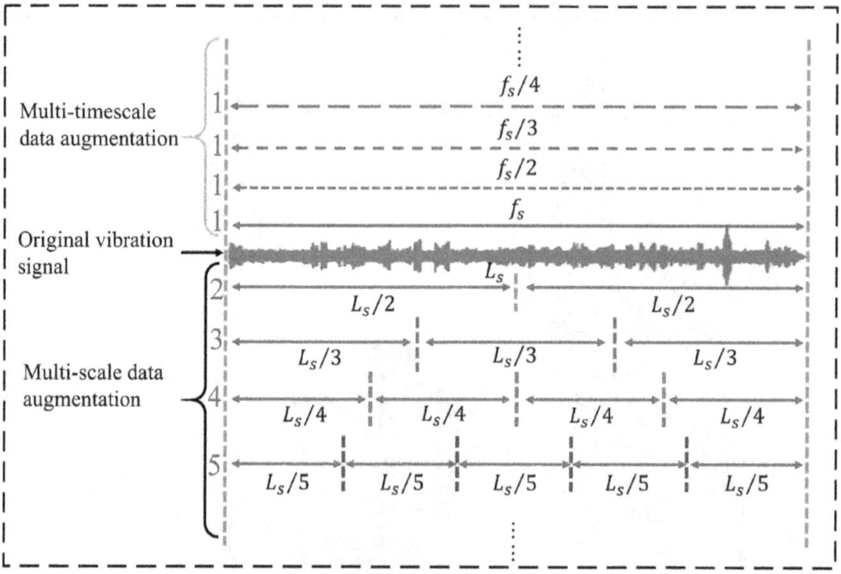

Figure 6.9 The data augmentation strategy diagram of the data used for IDDAN pretraining

transformer network, which includes one transformer encoder and a linear classifier. The transformer encoder is constructed by one multi-head attention module, one feedforward module and two normalization layers, which can automatically learn global features with the help of the SA mechanism [198]. A linear classifier is used to identify and distinguish bearing health conditions. As mentioned in Section 6.1, the SA mechanism can be processed parallelly with global capabilities, and long-distance information will not be weakened [199]. SA could be considered a CNN with a learnable receptive field. In other words, SA can learn receptive field automatically, but the receptive field of CNN needs manual adjustment and optimization of parameters.

As shown in Figure 6.10, the SA module includes a position-embedding module, a transformer encoder (feature extraction in Figure 6.8), and a linear layer. The linear layer can be regarded as a condition classifier to classify the global features extracted by the SA mechanism in the transformer encoder.

The motor signal of a certain length is first converted into the time–frequency map representation. For handling these 2D signal representations, the transformer encoder reshapes the image $x \in R^{H \times W \times C}$ into a sequence of flattened 2D patches $x_p \in R^{N \times (P^2 \cdot C)}$, where (H, W) is the resolution of the signal representation, C is the number of channels, and (P, P) is the resolution of each image patch. The effective input sequence length could be calculated through:

$$N = \frac{HW}{P^2} \tag{6.9}$$

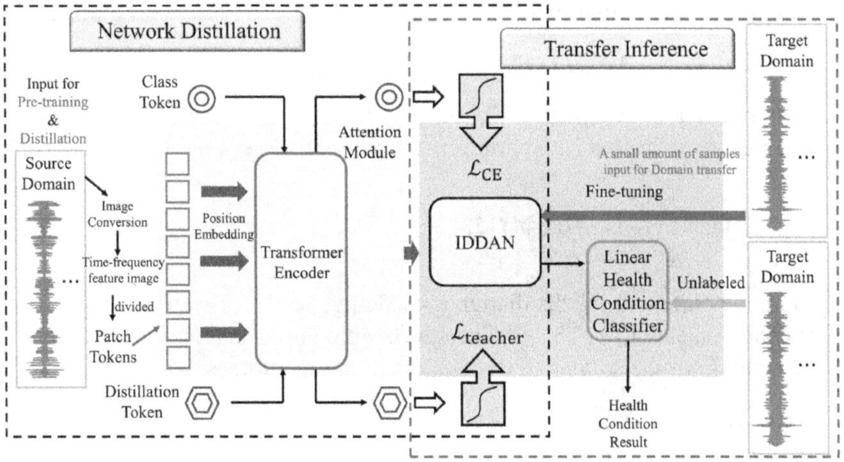

Figure 6.10 The structure presentation of IDDAN

where N stands for the resulting number of patches.

These patches representing machine health information have positions in the original time-series signal. Position embedding can achieve the effect of abstracting data in time series and represent relative or absolute position information in the input sequence. The implementation process of position embedding in this chapter is as follows:

$$z_0 = \left[x_{\text{class}}; x_p^1 E; x_p^2 E; \ldots; x_p^N E\right] + E_{\text{pos}} \tag{6.10}$$

where $E \in R^{\left(P^2 \cdot C\right) \times D}$, $E_{\text{pos}} \in R^{(N+1) \times D}$, E is the patch embedding projection, and D is the constant latent vector size through all of transformer layers.

The patches with marked positional information need to perform a normalization operation before entering the multi-head SA (MSA) mechanism. This chapter adopts layer normalization (LN) [200]. The state at the output of the transformer encoder z_L^0 serves as the image representation $y = LN\left(z_L^0\right)$. Both during pretraining and fine-tuning, a classification head is attached to z_L^0:

$$z_l' = \text{MSA}(LN(z_{l-1})) + z_{l-1} \tag{6.11}$$

where $l = 1, \ldots, L$. MSA is an extension of standard SA, which runs SA operations k times (the calculation principle has shown in Section 6.2.1):

$$\text{MSA}(z) = [SA_1(z); SA_2(z); \ldots; SA_k(z)]\, U_{\text{msa}} \tag{6.12}$$

For calculating MSA in this chapter, we set $SA(z) = Av$ and $U_{\text{msa}} \in R^{k \cdot D_h \times D}$, where $A = softmax\left(\frac{qk^{\mathrm{T}}}{\sqrt{D_h}}\right)$ ($A \in R^{N \times N}$, $D_h = \frac{D}{k}$ and $[q, k, v] = zU_{sa}$, $U_{sa} \in R^{D \times 3D_h}$).

As shown in Figure 6.10, LN is employed before both MSA and FFN blocks, and residual connections after every block. Our FFN block contains two layers: one hidden layer at pretraining time and a single linear layer at fine-tuning time:

$$\mathrm{FFN}(x) = (0, xW_1 + b_1)\, W_2 + b_2 \tag{6.13}$$

where two different parameters W_1 and W_2 from layer to layer are used, and a Gaussian error linear unit activation function is applied between layers [201]:

$$\left(1 + \tanh\left[\sqrt{\frac{2}{\pi}}(x + 0.044715x^3)\right]\right) \tag{6.14}$$

Both the class and the distillation embeddings of IDDAN are associated with linear condition classifier. The final health condition prediction result is determined by the addition of the *softmax* [202] outputs of the two routes:

$$Class_{\mathrm{predict}} = \mathrm{Linear}[softmax(O_{ce}) + softmax(O_{de})] \tag{6.15}$$

where O_{ce} and O_{de} are the output of the two classifiers, respectively.

Backbone network's distillation process is carried out simultaneously with pretraining. In the output of the *softmax* layer, in addition to positive examples, negative labels also carry a lot of information. This training method of KD makes each sample bring more information to student model than the traditional training method.

Firstly, in order to improve the recognition accuracy of health conditions, the teacher model is chosen as VGG-16. It is a strong feature extractor and classifier. In this distillation strategy, the hard decisions of the teacher model $y_t = \mathrm{argmax}_c Z_t(c)$ (Z_t is the logits of the teacher model) are true labels. The applied hard-label distillation objective should be defined as

$$L_{\mathrm{global}}^{\mathrm{HardDistill}} = \frac{1}{2} L_{\mathrm{CE}}(\psi(Z_s), y) + \frac{1}{2} L_{\mathrm{CE}}(\psi(Z_s), y_t) \tag{6.16}$$

where L_{CE} represents the cross-entropy, and Z_s is the logits of the student model (IDDAN). For the image represents of machine signal, it is possible that hard labels will change according to the DA (Section 6.3.1). In this hard-label distillation loss function, the teacher prediction y_t has the same role with the true label y.

Figure 6.10 demonstrates that the distillation, the class, and patch tokens interact through the SA layers in the transformer encoder. The output of the distillation token is the hard-label predicted by the teacher model. In this method, the class token w_{class} and the distillation token w_{distill} are trained by back-propagation algorithm:

$$w(m+1) = w(m) - \eta \frac{\partial J(w(m))}{\partial w(m)} \tag{6.17}$$

where $J(w)$ represents the training error at any instance, m denotes the number of iterations, and $\eta > 0$ is the preset learning rate before training. This chapter employs the gradient-based parameter optimizer AdamW, replacing L2 regularization of Adam with weight decay.

The transfer inference stage of IDDAN includes fine-tuning and condition classification. Fine-tuning is a pivotal step to infer the pretrained model to the target task through the transfer learning method. Condition classification is to use the inference completed mode to classify and diagnose fault conditions. The backbone network can be divided into feature extractors and classifiers. The feature extractor extracts the low-level features of the image. In the pretraining stage, the proposed vibration signal DA strategy can obtain large-scale pretraining samples of the source domain, and the pre-model trained with large-scale data has a higher generalization ability to extract the underlying features. Therefore, during the transfer process, the bottom layer weights are frozen, and the high layer weights are opened. In this chapter, the FFN layers and linear classifier layer in the pretrained IDDAN will be updated with parameters in fine-tuning (has been demonstrated in Figure 6.8). This is because the previous SA and LN layers are used to obtain a general representation from the image, and the latter FFN layer and linear classifier are more relevant to downstream special fault diagnosis tasks.

To verify this approach, here uses three professionally measured bearing datasets:

1. Dataset A: KAt-DataCenter bearing dataset [203] contains a variety of faults to perform fault diagnosis experiments. This dataset focuses on not only artificial bearing damages but also real damages. It could prove better than the proposed approach in this chapter is competent for different kinds of bearing fault diagnosis. The tested motor is a 425 W PMSM which has the nominal torque $T = 1.35$ N m, the nominal speed $n =$ 3,000 rpm, the nominal current $I = 2.3$ A and the number of pole pair $p = 4$.
2. Dataset B: The intelligent maintenance systems (IMS) bearing dataset is measured by and provided by the Center for Intelligent Maintenance Systems (IMS), University of Cincinnati. Recorded vibration signals include normal condition, rolling element fault, inner race fault, and outer race fault. Each data describes a test-to-failure experiment and consists of individual files that are 1 s vibration signal snapshots recorded at specific intervals. Each file consists of 20,480 points with the sampling rate set at 20 kHz and collected by NI DAQ Card 6062E.
3. Dataset C: Experiments of the CWRU bearing dataset were conducted using a 2 HP Reliance electric motor, and acceleration data were measured at locations near to and remote from the motor bearings. Motor bearings were seeded with faults using electro-discharge machining. Faults ranging from 0.007 in. in diameter to 0.040 in. in diameter were introduced separately at the rolling element, inner race, and outer race faults. Vibration data were collected at 12,000 samples per second, and data were also collected at 48,000 samples per second for drive end bearing faults. Speed and horsepower data were collected using the torque transducer.

These three bearing datasets include vibration signals, while they are obtained from different machines and different operation conditions. Their detailed information is displayed in Table 6.4.

Table 6.4 Detailed information of various experiment bearing datasets

Dataset names	Bearings	Conditions	Speed (rpm)	Load conditions
KAt	PMSM bearing	Normal	1,500	0.1, 0.7 N m
		Inner ring fault	1,500	0.1, 0.7 N m
		Outer ring fault	1,500	0.1, 0.7 N m
IMS	Rexnord ZA-2115 bearing	Normal	2,000	6,000 lbs
		Inner ring fault	2,000	6,000 lbs
		Rolling element fault	2,000	6,000 lbs
		Outer ring fault	2,000	6,000 lbs
CWRU	Motor bearing	Normal	1,750	0, 1, 2, 3 HP
		Inner ring fault	1,750	0, 1, 2, 3 HP
		Rolling element fault	1,750	0, 1,2,3 HP
		Outer ring fault	1,750	0, 1, 2, 3 HP

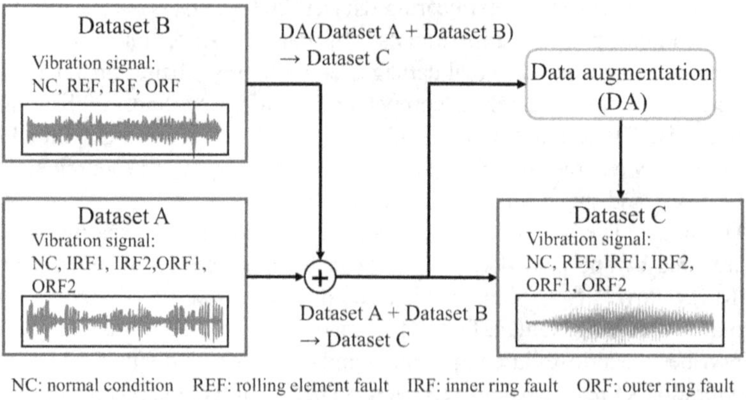

No.	IDDTN transfer experiments	Training data	Test data
1	Dataset A + Dataset B → Dataset C	100% labeled Dataset A + 100% labeled Dataset B + 20% labeled Dataset C	50% unlabeled Dataset C
2	**DA**(Dataset A + Dataset B) → Dataset C	DA(100% labeled Dataset A + 100% labeled Dataset B) + 20% labeled Dataset C	50% unlabeled Dataset C

Figure 6.11 Diagram of different transfer diagnosis experiments.

The planned transfer diagnosis experiment is shown in Figure 6.11, which includes the use of each dataset and the partition of training and test data. In this experiment, all used data are bearing vibration signals collected from different machines or devices. Therefore, this experiment put the target on testing the ability and performance of inferencing trained IDDAN to a new machine. It can be found that the datasets A and B are mixed as pretraining data. This is because the dataset A does not include vibration signal samples of rolling element fault. In No. 2 experiments, samples of two damage levels are also not included in the dataset B. No. 1 experiment tests the accuracy of the proposed method with standard data amount (2,000) which come from two different datasets, while No. 2 experiment tests with a larger amount of data (40,000). The percentage of the dataset means that the data are picked evenly from each fault and each working condition in the dataset. Overall, 20% of the target domain dataset is used in this section as the fine-tuning dataset.

We evaluate the proposed IDDAN through two designed experiments shown in Figure 6.11. In the first experiment, B refers to the source domain dataset and C refers to the target domain dataset. In the second experiment, A plus B turn into the source domain data, and C also refers to the target domain dataset. In each experiment, the training data include all the labeled samples from the source domain dataset, and the fine-turning data use 20% labeled data from the target domain dataset. Then, we randomly take 50% of unlabeled samples as the test data. The detailed information of training parameters is demonstrated as follows. The CPU and GPU devices for training are I9-12900K and RTX 3080Ti, respectively. The epoch for pretraining and fine-tuning is set to be 500 and 150, respectively. In addition, the patch size is set as 32, and the number of attention heads is 6. This step is trained using the AdamW optimizer with a learning rate 3×10^{-4}. In the fine-tuning step, the dimension of the MLP block is set as 2,048, the resolution of target domain time–frequency maps increases to 256 from 224. The training loss of experiments is plotted in Figure 6.12. As shown in Figure 6.12, the initial loss in the fine-tuning stage is significantly lower than in the pretraining stage.

The No. 1 experiment is repeated ten times with those preset parameters. In each experiment, all accuracies of the transfer diagnosis are over 82%, and the average accuracy is around 85%. For obtaining a baseline testing accuracy, we set a control transfer test using the CNN model to condition recognition. In this CNN module, the number of convolutional layers is 5, and the size of the convolution kernel is set to be

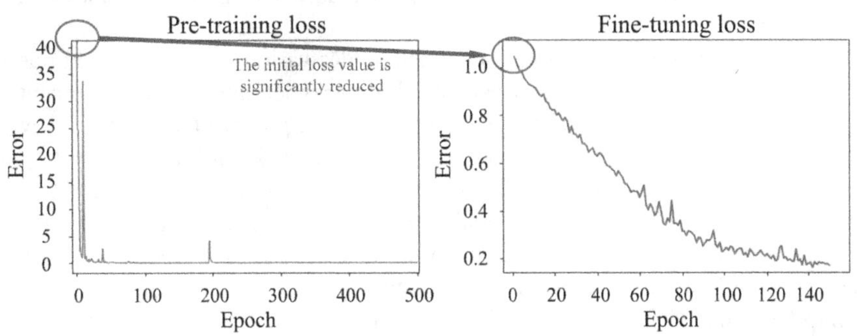

Figure 6.12 The pretraining loss (500 epoch) and the fine-tuning loss (150 epoch) curve

3. The five pooling layers are followed by each convolutional layer separately, and the size of the pooling kernel is 2. The CNN model is also pretrained and fine-tuned using the same data samples. According to the same experiments times with IDDAN, the average accuracy is around 89%, and the lowest accuracy is 85%. It means that the proposed IDDAN method can effectively diagnose the normal condition and three faults of bearing, but the accuracy of the CNN-based model is slightly higher than IDDAN with a small number of pretraining samples.

Due to the difference between the SA mechanism and the convolutional network, SA-based network structures benefit more from large-scale data [198]. The experiments of this part take the same settings as Figure 6.11 and the DA strategy is applied in the experiment. We use MATLAB® (signal processing toolbox) to perform the proposed DA method on all vibration signals in the source domain dataset, followed by the signal-to-image conversion. The samples used in the No. 2 experiment were expanded from 2,000 to 40,000, reaching 20 times the original pretraining samples.

In this section, the No. 2 experiments in Figure 6.11 should be repeated several times. In these experiments, the amount of data used in the fine-tuning stage as a percentage of the total target domain data (P_T) is kept as 20%. It can be found that all accuracies of the transfer diagnosis are over 92%, and the average accuracy is around 95%. To provide visual insights into the effects of features transferring from the source and target datasets, we use the *t*-distributed stochastic neighbor embedding (*t*-SNE) technique to map the high-dimensional features into a 2D space. Figure 6.13 demonstrates the feature recognition ability of the last layers before *softmax* layer when we completed the fine-tuning step.

When deploying the model in a device such as an EV, the floating point operations (FLOPs) of the method must be considered. FLOPs describe the computing power required by the deployed device, and the number of parameters describes the required memory size.

As shown in Figure 6.14, the IDDAN achieves a relatively minimum requirement for FOLPs and parameter quantities compared to some commonly used frameworks. In detail, the required FLOPs of IDDAN are 2.8GFLOPs and the number of parameters is 22.1 M, which is obviously less than the teacher model VGG-16.

The dataset of bearing vibration signals under various working conditions is shown in Table 6.2. Therefore, a very harsh transfer learning test environment is formed, which can well detect the performance of the proposed method. However, the number of labeled samples provided by the target domain is also the key to performance. We divide nine grades according to P_T from less to more: 1%, 5%, 10%, 15%, 20%, 25%, 30%, 40%, and 50%. The test data are still 50% unlabeled data of the target domain. It is necessary to compare the performance of the different methods. In this comparison, the historically outperforming classical algorithms SVM and CNN are replaced with IDDAN, and short-time Fourier transform and Hilbert–Huang transform will be replaced with CWT. DA is also added as a controllable condition. Results of all tests are summarized in Table 6.5, which shows that the proposed method can achieve the best performance if $P_T \geq 10\%$.

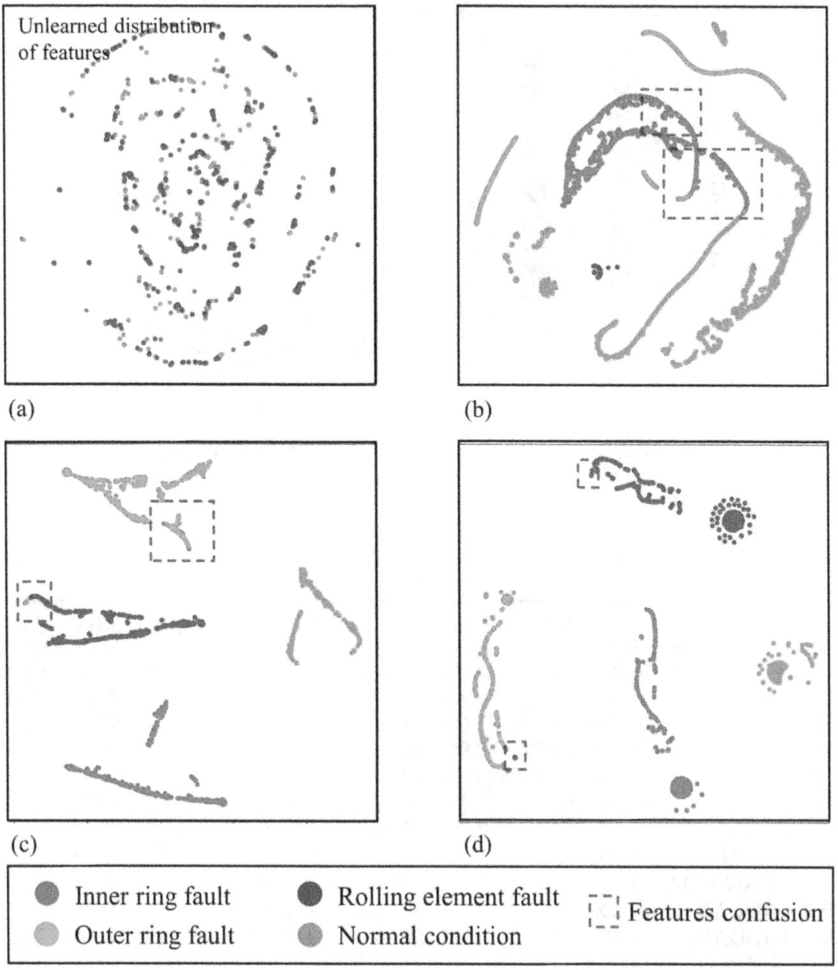

Figure 6.13 t-SNE visualization of features after fine-tuning (P_T = 20%): (a) before training; (b) small amount of pretraining samples; (c) CNN method with large amount of pretraining samples; and (d) IDDAN with large amount of pretraining samples

To demonstrate the performance of the proposed IDDAN, three different existing bearing diagnosis methods are used for comparison. Table 6.6 shows the comparison of the diagnostic results of other methods collected from the paper with the proposed IDDAN in transfer fault diagnosis experiments (P_T = 50%). All methods are tested on CWRU dataset.

The results demonstrate that fine-tune-based methods outperform all compared methods. The accuracy of the CNN model is better than IDDAN in experiments without DA. However, IDDAN overtakes the CNN model after going through DA

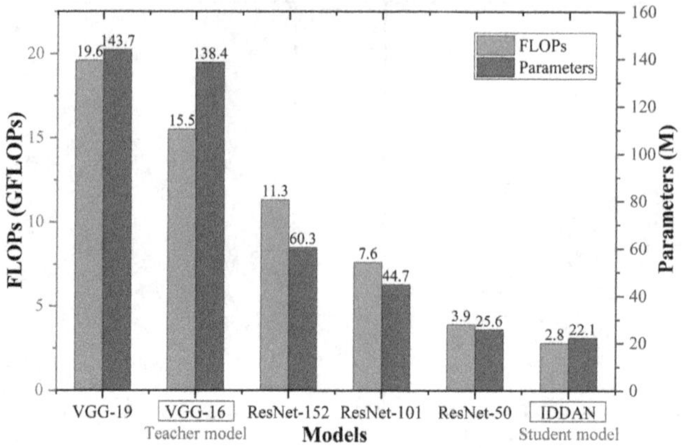

Figure 6.14 t-SNE visualization of features after fine-tuning (P_T = 20%)

Table 6.5 Result comparison

P_T Methods	1%	5%	10%	15%	20%	25%	30%	40%	50%
STFT + SVM	69.8%	73.9%	75.5%	77.2%	78.8%	82.7%	83.2%	87.3%	88.6%
STFT + CNN	67.9%	71.7%	78.4%	79.7%	85.3%	90.9%	91.0%	91.5%	92.4%
STFT + IDDAN	66.4%	66.8%	77.6%	78.1%	80.7%	84.4%	86.6%	93.2%	92.6%
HHT + SVM	71.6%	74.0%	80.9%	82.5%	80.4%	83.2%	81.5%	84.1%	89.8%
HHT + CNN	66.9%	75.5%	80.2%	84.6%	87.9%	91.3%	91.9%	90.7%	92.7%
HHT + IDDAN	60.1%	64.9%	77.1%	79.4%	84.2%	85.3%	90.6%	92.5%	93.6%
CWT + SVM	71.0%	79.4%	82.2%	83.7%	86.1%	87.8%	88.9%	90.5%	90.4%
CWT + CNN	69.9%	75.9%	80.3%	82.2%	89.5%	92.3%	92.2%	93.2%	93.9%
CWT + CNN + DA	71.4%	79.3%	86.7%	90.9%	94.1%	95.6%	95.7%	96.1%	96.2%
CWT + IDDAN	67.5%	69.2%	76.0%	80.5%	84.4%	87.9%	92.2%	92.7%	93.2%
CWT + IDDAN + DA	70.4%	85.5%	87.2%	92.2%	95.9%	96.7%	97.3%	99.0%	99.5%

Table 6.6 Fault diagnosis results compared with other existing methods (P_T = 50%)

Methods	Transfer method	Target domain dataset	Average accuracy (%)
CNN	Fine-tuning	CWRU	93.9
CNN-DA	Fine-tuning	CWRU	96.2
DDC [204]	Maximum mean discrepancy	CWRU	78.2
DCTLN [205]	Multiple domain adaptation	CWRU	86.8
DANN [206]	Domain adversarial	CWRU	80.9
IDDAN	Fine-tuning	CWRU	93.2
IDDAN-DA	Fine-tuning	CWRU	99.5

supporting. According to application conditions, it could be divided into two categories:

1. There are not any labeled data collected from the target domain machine. In this condition, domain adaptation-based methods are widely used for solving transfer learning problems. The most advanced research is based on the maximum mean MMD) of data samples. For example, DDC is to add an adaptation layer and an MMD module to the traditional CNN structure [204]. The DANN is the use of DNNs capable of domain discrimination [206]. The DCTLN add two domain adaptation losses to the CNN loss function to optimize the MMD distance between the source domain and target domain [205]. However, this method will not reach the highest accuracy when diagnosing faults in the target domain.
2. There are a few labeled data collected from the target domain machine. Fine-tuning is currently widely used in computer vision to further adapt to the target domain after pretraining. And the accuracy of the model after fine-tuning is often related to the amount of pretrained data. The application scenario that requires the target domain can provide a small amount of label data. At the same time, the SA mechanism included in IDDAN is proved to be more dependent on large-scale pretraining samples.

For further testing the ability of the proposed diagnostic method to identify fault features, we set two damage levels of inner and outer ring faults. In detail, the inner and the outer faults in CWRU bearing dataset are divided into 0.007 and 0.021 in., while the inner and outer faults in KAt bearing dataset are also subdivided into two severity levels: within 2 mm, between 2 and 4.5 mm.

The experiments are repeated several times with the proposed DA process in the pretraining stage. The result shows that faults recognition accuracies of the CNN-based transfer method and IDDAN are around 84% and 92% ($P_T = 50\%$). The *t*-SNE features visualizations are demonstrated in Figure 6.15. According to the information from parts C, D, and E, we can further compare and analyze the effectiveness of IDDAN and other methods. We can observe the following points:

1. Compared with the classical CNN based transfer learning method, IDDAN achieves higher classification accuracy when given enough pretraining data. This means that transfer learning-based diagnosis accuracy has been further improved. Its purpose is to accurately classify various faults of the target machine when the labeled data of the target machine are not enough to directly train the diagnosis model.
2. When classifying the inter ring fault and outer ring fault among the four fault types into two damage levels, the classification accuracy of the transfer learning-based diagnosis model decreases significantly. This is due to the high similarity between features of different damage levels. As can be intuitively seen in Figure 6.15, the classification model produces confusion between different fault types and levels.

6.2.2 Bearing fault diagnosis: example 2

Siamese neural networks are a class of neural network architectures that contain two or more identical subnetworks (Figure 6.16). The same network has the same configuration, including the same parameters and weights. Parameter updates are performed jointly on both subnetworks. Shared weights for subnetworks mean that training requires fewer parameters, less data and is less prone to overfitting. The framework was built to compare two images to determine similarity. The input to the framework is two images, and the output is a similarity value. In the training

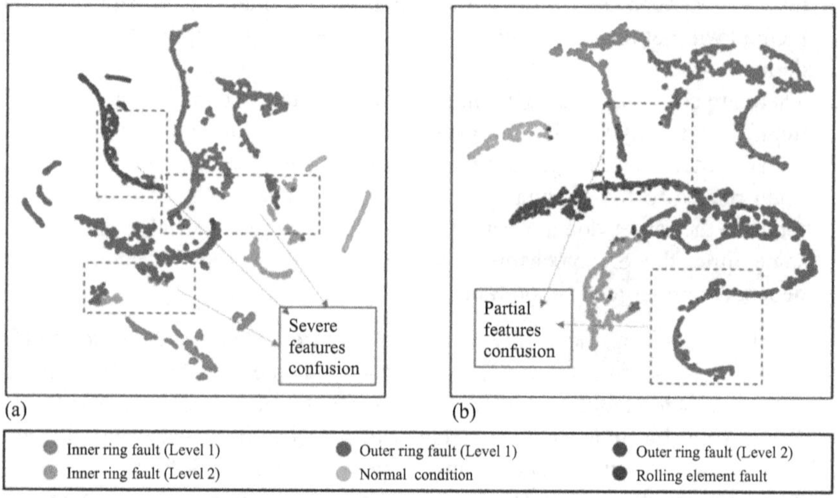

Figure 6.15 t-SNE visualization of features after fine-tuning (P_T = 50%): (a) CNN-based transfer method with large amount of pretraining samples and (b) IDDAN with large amount of pretraining samples

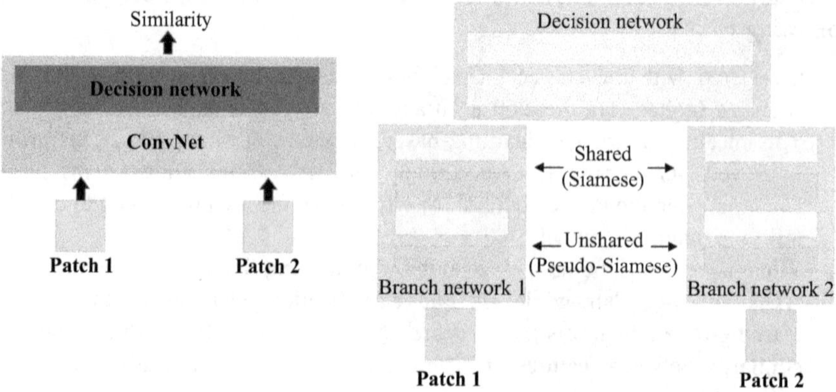

Figure 6.16 Framework of Siamese neural networks

data used in the literature, the output value is labeled as $y = 1$ if the two images match, and if the two images do not match then the training data are labeled as $y = -1$. This training data are labeled in such a way as to obtain the matching value. Patches 1 and 2 are passed through a two-channel network for extracting feature vectors separately. The feature extraction process for the two images patches 1 and 2 is independent of each other. A similarity loss function is then done on the two feature vectors at the decision layer for network training. As shown in the Figure 6.17, the input is viewed as a two-channel image. Two (1, 64, 64) single-channel data are synthesized into a (2, 64, 64) two-channel matrix, and the matrix data are then used as the input to the network. The explicit feature extraction process of branching is skipped, and instead the similarity evaluation function is learned directly. The last layer is directly a fully connected layer with the number of neurons as 1, which directly represents the similarity of the two images. If the input is a two-channel image, which is equivalent to a network with two feature maps, after the first layer of convolution, the pixels of the two images are combined and mapped in a relevant weighted way.

The attention mechanism in neural networks is a resource allocation scheme that allocates computational resources to more important tasks with limited computational power, while solving the information overload problem. In general, the more the parameters a model has, the more the expressive it is. At the same time, the amount of information stored in the model can lead to information overload. The attention mechanism can improve the efficiency and accuracy of task

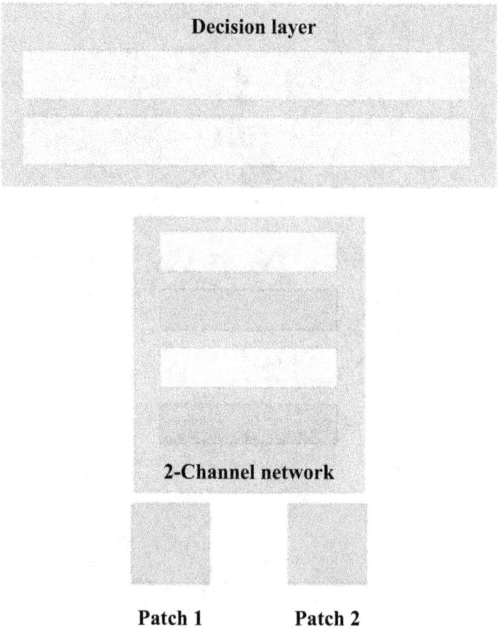

Figure 6.17 The frame of two-channel network

processing by focusing on the information that is more critical to the current task in the input information, reducing the attention to other information, or even filtering out irrelevant information.

The attention distribution uses the input information vector as the source of information and a query vector to find and select the relevant information in the input signal. A "soft" selection mechanism is adopted, where relevant information is extracted from all information, and the features with the highest relevance coefficients are given the most weight. The attention distribution α_i indicates how relevant the ith piece of information in the input information vector X to the query q. A "soft" information selection mechanism is used to give the results of the query, which is a weighted average of the input information, to obtain the attention value:

$$\text{att}(X, q) = \sum_{i=1}^{N} \alpha_i x_i \tag{6.18}$$

The attention mechanism is an addressing operation that determines the output by calculating the similarity of the Query to the address Key of the element within the input signal (Figure 6.18). The similarity of the Query to the Key is used to calculate the weight of each value, and then the value is weighted and summed. This is equivalent to calculate the weight of each value from the similarity of the Query and Key, and then weighting the value to obtain the final value, which is the Attention value.

Step 1: Calculate the similarity between Query and Key:

$$S_i = F(Q, K_i) \tag{6.19}$$

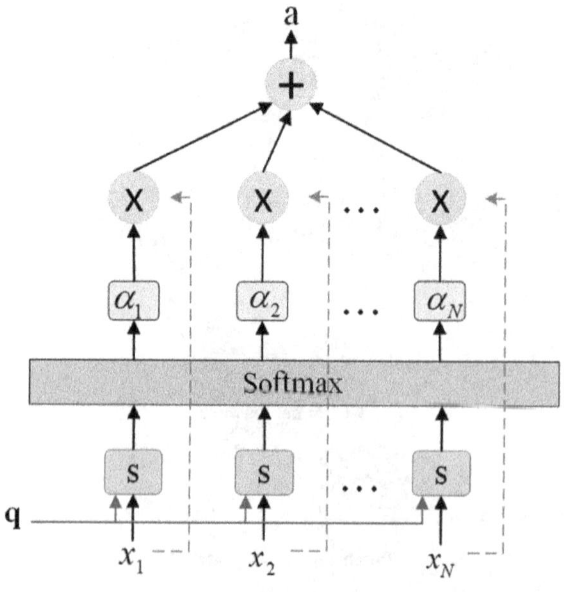

Figure 6.18 Diagram of attention mechanism

The attention scores are then transformed numerically by means of the *softmax* function. A probability distribution with all weight coefficients summing to 1 is obtained:

$$\alpha_i = softmax(S_i) = \frac{\exp(S_i)}{\sum_{j=1}^{N} \exp(S_j)} \tag{6.20}$$

Finally, the value is weighted and summed according to the weighting factor:

$$\text{Attention}((K, V), Q) = \sum_{i=1}^{N} \alpha_i v_i \tag{6.21}$$

This chapter presents a dual-channel Siamese network based on an attention mechanism for the diagnosis of bearing faults. The proposed method uses a self-attentive mechanism-based module for signal feature extraction and then diagnoses the fault type by calculating the Euclidean distance of the signal output of the dual channels. Commonly used fault diagnosis algorithms usually take the extraction of local features of the fault signal. The proposed method can effectively focus on global faults and achieve high diagnostic accuracy. In a supervised learning paradigm, the proposed method maximizes the representation of different labels and minimizes the representation of the same label. In addition, the proposed method is suitable for one-shot learning and is independent of false samples, so it can be used for bearing fault diagnosis problems that require a strict fault-tolerant process.

The input is mapped as a feature vector, and the difference between the inputs (the semantic gap in the image) is represented using the distance between the two vectors $L1$ Norm. Two samples at a time are needed as a sample pair to calculate the loss function:

$$E_w(X_1, X_2) = \|G_W(X_1) - G_W(X_2))\| \tag{6.22}$$

The Siamese network has two subnetworks with the same structure and shared weights. The two inputs X_1 and X_2 are received separately, converted into vectors $G_W(X_1)$ and $G_W(X_2)$, and the distance between the two output vectors is calculated by means of some distance metric E_w:

$$\text{Loss} = \frac{1}{2N} \sum_{n=1}^{N} \left[YE_w^2 + (1 - Y)\max(0, m - E_w)^2 \right] \tag{6.23}$$

The previous loss function is mainly used in dimensionality reduction, where similar samples remain the same in the feature space after feature extraction. When $Y = 1$, the minimization loss function makes the sample distance of similar samples smaller. When $Y = 0$, the loss function is maximized and the sample distance of the dissimilar samples increases. The sample type is identified by determining whether the sample distance exceeds the threshold.

In this section, the benchmark motor bearing failure dataset used in this study is first described. The experimental setup used to evaluate the proposed bearing fault severity classification framework will then be presented. The proposed algorithm is used in the bearing fault dataset, and classification accuracy results are obtained and compared with other classical algorithms for validation.

The test rig consists of a 1.5 kW motor, torque transducer, power test meter, and electronic controller. The bearings to be tested support the rotating shaft of the motor, the drive end bearing is SKF 6205, and the fan end bearing is SKF 6203. The geometrical dimensions of the two bearings and the frequency of failures seen in each part are listed in Table 6.7. An acceleration sensor is placed above the bearing housing on the fan and drive side of the motor to collect the vibration acceleration signal of the faulty bearing. The vibration signals were acquired by a 16-channel data logger with a sampling frequency of 12 kHz, the drive-side bearing failure also included data with a sampling frequency of 48 kHz. The dataset is a human-made fault with obvious characteristics and relatively easy to diagnose. The dataset is widely used and well recognized. It can be used as a base dataset for algorithmic testing (Figure 6.19).

In this chapter, the proposed model is optimized by adjusting four hyper-parameters to achieve the best diagnostic performance. The number of stacked layers of the model, the learning rate, the number of transformer heads, and the optimization algorithm. The experimental results are shown in Table 6.8. Experimental results indicate that the lack of sufficient depth in the model may fail to obtain low-dimensional features of faulty samples. However, when the model depth is stacked too much, the training speed of the network will be slow, resulting

Table 6.7 Bearing fault types

Fault type	Ball defect (0.007 in.)	Ball defect (0.014 in.)	Ball defect (0.021 in.)	Inner race fault (0.007 in.)	Inner race fault (0.014 in.)
Label	1	2	3	4	5
Fault type	Inner race fault (0.021 in.)	Normal	Outer race fault (0.007 in.)	Outer race fault (0.014 in.)	Outer race fault (0.021 in.)
Label	6	7	8	9	10

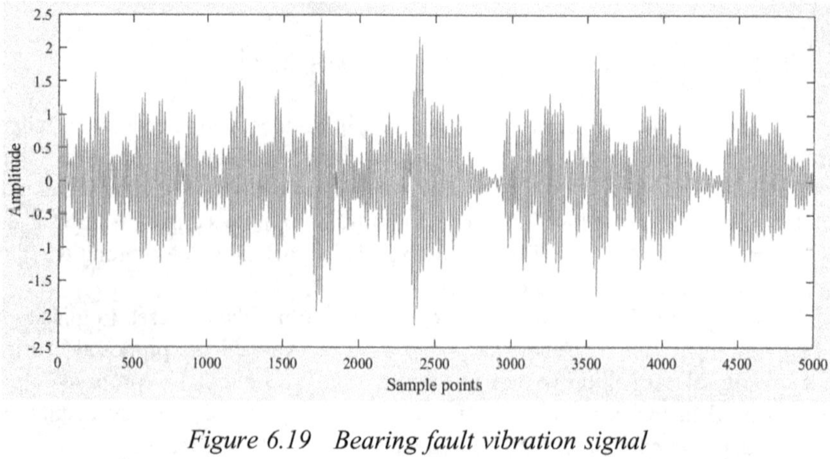

Figure 6.19 Bearing fault vibration signal

Table 6.8 Accuracy comparison

Depth	Learning rate	Number of head	Optimization algorithm	Accuracy (%)
8	0.01	4	Adam	98.64
16	0.01	4	Adam	96.17
4	0.01	4	Adam	95.43
8	0.01	4	Adam	98.64
8	0.05	4	Adam	97.45
8	0.1	4	Adam	90.12
8	0.01	2	Adam	96.34
8	0.01	4	Adam	98.64
8	0.01	8	Adam	92.37
8	0.01	4	Adam	97.15
8	0.01	4	SGD	94.68
8	0.01	4	AdaGrad	98.64

in difficult convergence and suboptimal fault diagnosis rate. Therefore, the maximum number of depths of the proposed model is 8 to achieve the best diagnostic performance. Comparing the fault diagnosis accuracy for different stacking numbers shows that a depth of 8 performs the best in training and diagnosis. This can be explained that under a certain convergence performance, the higher the depth of the model, the stronger the ability to extract signal features. Therefore, in subsequent experiments, the depth of the model was set to 8, and then three other model parameters were adjusted. In addition, the learning rate determines the step size of the weight iteration, and its impact on the model performance is reflected in the size of the initial learning rate and the speed of the learned transformation-based model. The initial learning rate needs to be continuously adjusted to the optimal value for diagnostic efficiency. A larger learning rate can cause the model to not converge and vice versa can cause the model to converge slowly or fail to learn. Furthermore, as the learning rate increases, the model changes from underfitting to overfitting. Comparing the accuracy of different learning rate states shows that the model achieves the best performance with a learning rate of 0.01 compared to other learning rates. The multi-head attention in the model is a combination of multiple SAs. Instead of computing each head cyclically, multi-head is implemented using matrix multiplication through transposition and shaping. By splitting the dimensionality, the model performance is improved without increasing complexity. The best performance was obtained in experiments with four heads compared to a different number. Furthermore, when the training data N is very large, it is difficult to compute the total cost function to find the gradient, and SGD is a common method for computing the mini-batch in the training set. However, progress along the shallow dimension is very slow and jitters with steep direction. Similar to learning rate annealing over time, AdaGrad is a parameter-based adaptive learning rate method with different learning rates for different parameters. However, the disadvantage of AdaGrad is that the monotonic learning rate is too aggressive and stops learning too early. As Adam's fault diagnosis accuracy outperforms the other

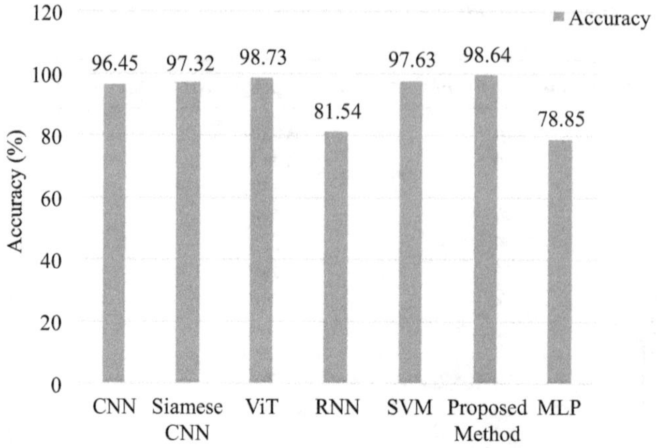

Figure 6.20 The diagnosis performance of different models.

four optimization algorithms under the same conditions, it was chosen as the optimal algorithm for the network.

In this chapter, the proposed algorithms are tested against traditional algorithms to verify the superiority of the proposed algorithms. All algorithms were applied to the same dataset and optimized for the structure. The experimental results are shown in Figure 6.20. First, as a multilayer supervised learning neural network, the CNN extracts the input features through convolutional filters to achieve efficient feature extraction to obtain well classification performance. An accuracy of 96.45% was achieved in bearing fault diagnosis. Compared to CNN, the Siamese CNN model with a dual-channel Siamese neural network architecture achieves better classification performance at 97.32%. It shows that weight sharing amplifies the dimensional characteristics of feature extraction and gives better separation results. In addition, the attention-based ViT model achieved the fault classification accuracy of 98.73%. In addition, the RNN neural network based on recursive calculation of the system state achieved a fault diagnosis accuracy of 81.54%. Compared to the basic feed-forward neural network model MLP and the binary classification model SVM, the proposed method achieves a fault diagnosis accuracy of 98.64%. Thus, in comparison with other conventional algorithms, the proposed method can extract global fault features and achieve superior classification performance.

In summary, this chapter reviews the application of AI techniques in motor fault diagnosis. Comprehensive studies of motor faults and their severity are still rare and have so far been limited to diagnosing faults in motors under specific operating conditions. It is difficult to detect faults in motors under light loads. In addition, the accuracy of fault diagnosis may be reduced due to fluctuations in rotor speed during data acquisition at different loads. Therefore, it remains an open challenge to consider the impact of motor operating conditions on AI-based fault diagnosis. AI for pattern recognition or fault diagnosis includes a large number of different types of mathematical tools, that is, preprocessing, extraction and selection of appropriate statistical features, and selection of model parameters. It is a challenge to choose which tool is best suited for a particular problem and machine in different situations. Moreover, there is a lot of scope for research in this area as condition monitoring techniques for fault detection and diagnosis of rotating machinery have been improved from traditional methods to AI methods. AI-based diagnostic systems still have some challenging tasks to accomplish in terms of their efficiency, reliability, computation time, adequate database, and robustness.

Chapter 7

Case study 4—AI-based gearbox fault diagnosis

Generally, in the electric powertrain system of a machine, transmission simply refers to the gearbox that uses gears and gear trains to provide speed and torque conversion from a rotating power source to another device. The gearbox is widely used in various rotating machineries, such as wind turbines, power generation devices, particularly EV. Gearbox failures account for a large proportion of the total failure time of all the statistics [207], and the initial fault features of the gearbox are weak and hidden in a set of time-varying vibration signals, which are difficult to effectively extract under strong noise. Most transmission system gearbox fault diagnosis methods are based on vibration and current methods (VA and MCSA). Among them, VA needs to install a sensor in the gearbox, while MCSA is more stable and saves costs. In vibration-based fault diagnosis, vibration signals are often accompanied by Gaussian and non-Gaussian background noise, and it is not easy to detect weak faults. The new strategy of LEASgram gearbox fault feature extraction proposed by He *et al.* [208] overcomes this problem. Experiments show that under the condition of Gaussian and non-Gaussian, interference and the resonance frequencies of multiple faults are easily coupled with each other, and the LEASgram method has significant fault identification and separation capabilities. In recent years, there are many new feature strategies partially based AI technologies designed for gearbox such as multiclass relevance vector machine (mRVM) [209], multiview sparse filtering (MVSF) [210], deep recursive dynamic principal component analysis (Deep RDPCA) [211], resonance residual technique [212], and deep learning–based methods [213,214].

In the application field of ML, Li [215] introduced a multimodal deep support vector classification (MDSVC) approach to complete the fault diagnosis task of the gearbox in which support vector classifier (SVC) is used to fuse Gaussian–Bernoulli deep Boltzmann machine (GDBM) in different ways to build MDSVC model. Among the classic ML classifiers, SVC- and SVC-derived classifiers are valued by researchers due to their excellent performance. Reference [216] proposed a gearbox fault detection method based on wavelet support vector machine (WSVM) and immune genetic algorithm (IGA). There are also studies using support vector machines. In [217], the optimized *k*-NN algorithm is used to identify the severity of gear cracks under different motor speeds and loads. Reference [218] also applies *k*-NN-based algorithm to gear fault identification. Decision tree (DT), as a basic classifier, can convert complex gearbox failure data into simple decision-making processes to extract rules [219].

Deep learning, which can analyze fault features deeper than classic classification algorithms, has also achieved great success in the classification task of gearbox. BP neural network is a classic and widely used neural network. Cheng *et al.* [220] used the IQPSO algorithm based on the PSO algorithm to optimize the classic BP neural network, and the IQPSOBP obtained after training has a higher accuracy for gearbox failures than PSOBP. In [221], the author compares the fault detection performance of the classic ML algorithms with the proposed two-dimensional (2-D) CNN architecture (Figure 7.1). The input data is got from multi-sensor fusion information. This 2-D

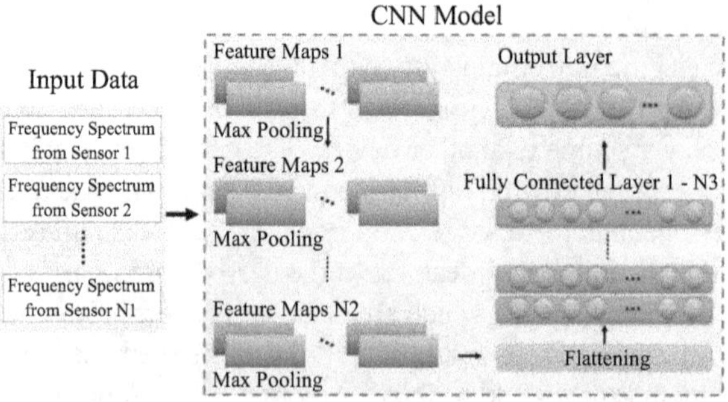

Figure 7.1 Diagram of CNN fault diagnosis architecture for dealing with multi-sensor fusion information (N1=numbers of sensors, N2 = numbers of convolutional and pooling pairs, N3 = numbers of fully connected layers; the output layer is based on a SoftMax function)

Table 7.1 Summary of contribution and AI algorithms of each reference with signals and feature methods

Reference	Signal	Feature method	AI algorithm	Contributions
[221]	Motor current	FFT	2-D CNN	• Data and feature level fusion • To preserve the dependency between different sensors
[222]	Vibration	MSCNN	MSCNN	• To introduce a multiscale coarse-grained layer into the traditional CNN • To learn discriminative features automatically and directly from raw temporal vibration signals • To classify different conditions simultaneously
[223]	Vibration	FFT	CNN	• The eigenvector is composed of RMS value, standard deviation, skewness, kurtosis, rotation frequency, and applied load

(Continues)

Table 7.1 (Continued)

Reference	Signal	Feature method	AI algorithm	Contributions
[215]	Vibration	GDBM	MDSVC	• To use GDBM to learn time, frequency, and pattern representation of features in wavelet modals
[216]	Vibration	EMD	IGA-WSVM	• To obtain the best WSVM parameters by IGA
[217]	Vibration	WPT, PCA	*k*-NN	• To identify five types of gear crack levels
[218]	Vibration	HT and FT	Weighted *k*-NN	• To use two-stage feature selection and weighting technique TFSWT technology to select sensitive features and remove irrelevant features • To identify early cracks in gears
[209]	Vibration	–	mRVM	• To design two features: accumulative amplitudes of carrier orders (AACO) and energy ratio based on difference spectra (ERDS)
[219]	Vibration	–	DT	• Calculation process is simple and fast by using DT
[210]	Motor current	MVSF	SVM	• To implement easily • To propose MVSF for integrating multi-view feature representation effectively
[224]	stator or/ and rotor current	–	multiclass SVM	• To define different time domain and frequency domain features of gearbox faults in DFIG stator and rotor current signals • To use decision-level information fusion for different solutions
[225]	Rotor current	HT	AE, SVM	• To design angle resampling algorithm • To design and apply a new type of deep structure classifier
[226]	Motor current	–	Radial basis function kernel-SVM	• To be used under variable speed conditions • To integrate an adaptive signal resampling algorithm, a frequency tracker, and a feature generation algorithm

neural network analyzes the current signals of multiple current sensors in real time and can be directly used to classify fault types, partially eliminating the need for complicated processing procedures in the middle, while a new multiscale convolutional neural network (MSCNN) [222] is proposed to be used to identify the health status of gearboxes. It can be used for both feature extraction and fault classification. Reference [223] also recommends the use of CNN to diagnose gearboxes.

Chapter 8

Discussion and future development of AI-based fault diagnosis for electric powertrain in EVs

As far as fault diagnosis is concerned, the two most important steps are how to extract fault features from a large amount of data and fault classification algorithms. Accurate extraction of fault features is the key to whether the fault can be classified correctly. It is the key to predict the occurrence of the fault in advance to be able to extract the fault characteristics from the weak fault signal when the fault first appears. The current diagnosis methods focus on the application and innovation of feature methods, while less attention is paid to the new application of AI algorithms. AI classification algorithms are also related to the speed and accuracy of fault classification, and the realization of multi-fault identification also relies on machine learning (ML) and deep learning. This chapter provides a brief description of the application trend of main AI algorithms.

1. *K*-NN
   The *K*-NN algorithm is widely used in the research of fault diagnosis due to its simple structure and can be used as a classifier of fault features.
2. Classic ML classifiers (SVM, RF, etc.)
   SVM has always been a very classic classifier, with good classification effect and fast calculation speed. The RF algorithm is a representative algorithm for ensemble learning, with excellent classification effects. If the training dataset is small, the effect of these classifiers is likely to be due to the deep learning classifier based on neural networks. But in recent years, compared with the diagnosis research based on neural network, the fault diagnosis research based on this kind of algorithm is less.
3. ANN
   Fault diagnosis models based on various ANN variants were popular in the early years and often appeared in combination with various feature extraction methods such as WT. In this type of diagnostic model, ANN takes various features extracted from the original signal as input and outputs the classification result. In addition, the running speed of the ANN algorithm is slower than that of the classic ML algorithms in part A and B, but it has a greater advantage if the amount of training data is large.
4. CNN
   CNN has got the great interest of people in recent years, which is reflected in the accelerated growth of various CNN-related chapters on fault diagnosis.

This is because CNN has both powerful automatic feature extraction and fault classification abilities, which is able to make the diagnostic model more integrated. One-dimensional CNN is mainly used to process the signals collected by the original sensors. The two-dimensional CNN is more used in combination with signal-to-image conversion methods, which helps to show the two-dimensional correlation between fault features.

5. RNN and long- and short-time memory (LSTM) and GRU
RNN is good at processing time series data and can predict system parameters and indicators from historical time data. Due to standard RNN algorithms often suffer from gradient disappearance or gradient explosion problems, the model cannot be trained normally. In recent years, LSTM has been more used in prediction tasks, such as RUL prediction for electric motors and battery systems, and SOC and SOH prediction for battery systems. GRU has a more forget gate than LSTM but has fewer parameters than LSTM, so the training speed is faster. It usually performs better than LSTM on the dataset with a small amount of data.
6. Cluster
Clustering algorithms, such as *K*-means and fuzzy *K*-means, often have two functions. On the one hand, in unsupervised or self-supervised diagnosis models, clustering algorithms can classify feature vectors with similar features into one category. On the other hand, clustering algorithms also have a certain feature dimensionality reduction effect.

AI-based techniques have been proven to be highly effective for fault detection and diagnosis in motor drive systems. Deep neural network-based methods have gradually become attractive in industrial applications due to the continuous upgrading of computer hardware. As the layers of neural networks deepen leading to the disappearance of locally optimal solutions and gradients. Pre-training methods alleviate the local optimal solution problem, and deep learning gradually becomes popular, including DBN, CNN, RNN, and LSTM. To overcome gradient disappearance, transfer functions, such as ReLU and max out, replace sigmoid and form the basic form of DNN. DBN simply converts an image matrix into a one-dimensional vector as input without considering the two-dimensional structure information of the image. It can represent the distribution of data from a statistical point of view and can reflect the similarity of similar data itself. However, the generative model cannot obtain the optimal classification surface between different categories, resulting in the classification accuracy may not be as high as the discriminative model. In the area of image recognition, the convolutional neural network is a popular study topic. Its weight-sharing network topology resembles biological brain networks, minimizing the complexity of the model, and the number of weights. Traditional recognition methods require extensive feature extraction and data reconstruction, so the image may be utilized directly as the input of the model. The signal of each layer of neuron in CNN can only propagate up one layer, and the processing of samples is independent at each moment, called forward neural network. In RNN, the output of a neuron can act directly on itself at the next timestamp. In addition to solving the temporal gradient disappearance, LSTMs are proposed to prevent gradient disappearance by implementing temporal memory function through gate switching. GAN is a generative learning algorithm that can efficiently mine the intrinsic distribution of a dataset using a small number of samples. In addition, the discriminative power of GAN discriminators makes it possible to build integrated fault diagnosis models. Besides, autoencoders can perform unsupervised learning from data samples and achieve great performance on different datasets without any new feature engineering (Table 8.1).

Most of algorithms are supervised learning-based approaches, which usually require manually designing training datasets in practical applications. However, the application of unsupervised learning in health monitoring needs further research. AE is a classic unsupervised feature learning method, but transfer learning is often needed to apply the trained model to other applications and additional training processes are needed to adjust parameters. Self-supervised learning is also a feasible research direction. As in Reference [276], unsupervised learning has been used in the field of fault diagnosis. It is not simply extracting fault features from historical fault signals, and then using labels for supervised classification. Instead, algorithms are used to detect the severity of multiple faults under variable speed and load online.

At present, machine health monitoring in engineering has applied a variety of algorithms such as ML and deep learning, but it has not followed the development of advanced algorithms, such as dynamic neural networks and the very popular transformer structure in natural language processing (NLP). The emergence of these cutting-edge algorithms can also have an impact on AI-supported health monitoring, but the application of these new frameworks in engineering is lagging behind. It should be that few people can transfer algorithms that are not yet fully mature in computer science to the field of engineering. For example, the attention technology in the transformer model, and so on, applied to fault diagnosis or even engineering, may produce better results.

Table 8.1 Highlight of other AI-based methods

Reference	Theme	Principle	Highlight
[117]	ANFIS	• If-then rule • Membership functions input	• Nonlinear approximation capability • High adaptability
[118]	DTANFIS	• Rule conversion • Vibration and current signal input	• To reduce data volume • To enhance classification performance • Less accurate classification of the current signal
[119]	ASSBDIM	• Sparse filtering	• High accuracy • Low calculation cost

(Continues)

Table 8.1 (*Continued*)

Reference	Theme	Principle	Highlight
		• Singular spectrum analysis • Time series data input	• Suitable for online applications
[59]	HMM-ANFIS	• Markov model prediction • Probability density function	• Nonlinear mapping • Real-time state estimation
[120]	Compress data-Deep learning	• Nonlinear projection for compression capture • Unsupervised learning	• Do not rely on a priori knowledge • To dig hidden data • Suitable for handling large amounts of data
[136]	Feature incremental broad learning	• Reconfiguration system • Simplify the model	• High diagnostic accuracy • Less training time • The feature extraction method is not optimal
[126]	Infrared Thermography ML	• Two-dimensional thermal image input • Optimal eigenvector matrix transfer	• Nondestructive • Non-contact detection • Performance affected by redundant information noise
[127]	Transfer learning	• Feature transfer • High-dimensional mapping of data	• To reduce feature data requirement • Cannot be applied to multi-target working condition
[128]	TFR	• Smooth blurred plane feature extraction	• To reduce computational burden • To improve feature separability
[121]	GA-PCA LDA ANN	• Estimating high-dimensional hybrid feature sets	• Simultaneous monitoring multiple faults
[28]	GA-ANN	• Weighting and threshold optimization	• Fast structure determination • High classification performance
[134]	AAS	• Unsupervised classification • Data clustering	• To improve monitoring effectiveness • Removal of redundant or irrelevant information
[129,133]	GA-AIS	• Spatial transformation	• To reduce computational complexity
[135]	Faster AP-MSDL	• Adaptive parameter estimation • Multi-scale learning	• Short running time • Great fault detection performance

The detection of a single fault in EV powertrains can no longer meet the needs of people. In the application of actual fault detection, it is often necessary to detect more and more fault types. Therefore, accurate and rapid multi-fault diagnosis is also an important direction of current development. We can adopt the thinking of lifelong learning/continuous learning in computer science. Using traditional ML classification algorithms and neural networks for multi-fault detection problems is not always having good accuracy/performance in distinguishing each fault. The difference between lifelong learning and transfer learning is that transfer learning will forget the content learned in previous tasks after learning new tasks, and continuous learning will not forget the tasks learned earlier and will make the trained model perform well in all tasks as possible. In practical applications, the update of this algorithm does not require the participation of old data, and the model in use can be quickly updated to achieve the purpose of extending the function of the existing algorithms.

Most AI-based data-driven methods applied in EV powertrains can achieve high-precision prediction and fault diagnosis without prior knowledge and physical model knowledge. On the other hand, many model-based conventional methods can build more accurate physical models by the intervention of AI.

In EV applications, most of the existing AI-based fault diagnosis methods are supervised learning methods. The supervised learning method is characterized by the need to use a large amount of labeled data for AI model training. In particular, the data labeling work that relies on the method of manually extracting the fault features from signals requires a lot of manpower and time. Although the trained model can achieve higher accuracy, it is not cost-effective in actual EV applications.

At present, there have been a few studies applying transfer learning to solve health monitoring problems in EV powertrains [191]. Using transfer learning can make the AI model transfer knowledge from the existing data to reduce or even eliminate the need for training data of the machine to be tested. Using unlabeled data can directly detect failures.

Self-supervised learning methods can also solve the expensive and time-consuming data labeling problem. Figure 8.1 demonstrates the diagram of contrastive self-supervised learning for condition classification. For example, in [276], more actions of unsupervised learning had been taken than simply feature extraction from historical fault signals and supervised classification. It detects the severity of many types of faults with various motor speed and conduct data analyses with unlabeled data. Interestingly, this method can detect the faults of three different components in the powertrain at the same time (Figure 8.2) and can diagnose the severity of each fault.

From this example, it can be analyzed that another AI application prospect is to propose a single model for multi-component, multi-fault, and multi-severity diagnosis. Both the existing AI-based methods and conventional methods can only diagnose the faults of a single component, or even single fault. In order to monitor and diagnose the health status of the entire EV powertrain, multiple methods or models are often needed. Therefore, achieving the single-model full diagnosis can greatly reduce the burden of on-board computing devices.

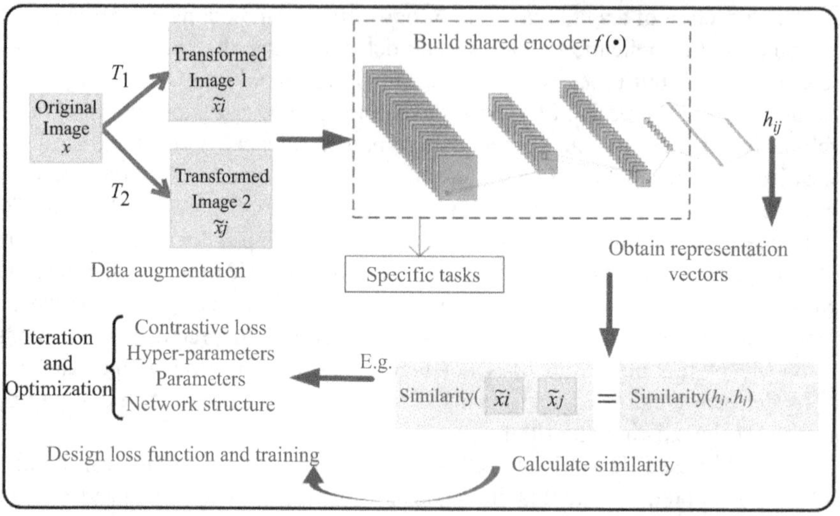

Figure 8.1 Flowchart of the contrastive self-supervised learning diagram

Figure 8.2 Faulty components of electric powertrain: small gear tooth damage (left), outer-race damaged bearing (mid), and 10% inter-turn short-circuit fault in the stator (right) [276]

The EV industry is rapidly evolving, and new technologies, components, and architectures are being introduced. Transferring diagnostic models to new generations of EV powertrains requires continuous updates and adaptations to accommodate new features and fault characteristics. The training of the EV fault diagnosis model is very time-consuming. Aiming at the problem that EV companies need to frequently update and upgrade their condition monitoring and fault diagnosis models, here are two application prospects and corresponding

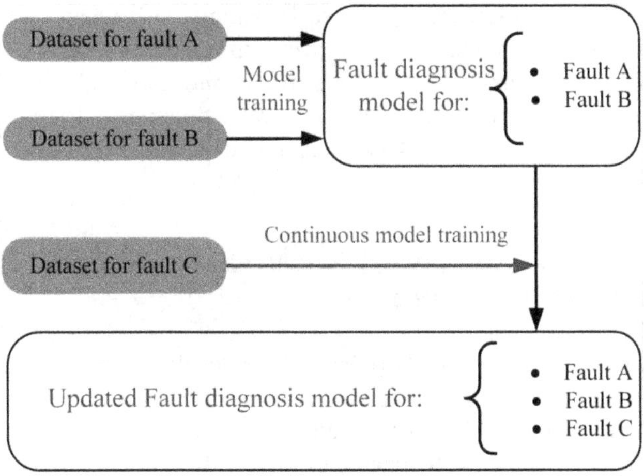

Figure 8.3 Schematic diagram of quickly updating and upgrading the model

AI technologies: Quickly updating and upgrading the model and quickly rebuilding the model:

1. Quickly updating and upgrading the model.
   Continuous learning technology is applied to the fault diagnosis model of EV powertrains, so that it can add new diagnosis tasks to the previously trained model (Figure 8.3). In particular, by combining IoT and cloud computing, it can be achieved that quick updates and upgrades of the model in EV powertrain safety applications. This approach significantly reduces the time and cost associated with traditional model updates and upgrades. However, considerations, such as data privacy, network security, and compliance, need to be addressed, and the updated and upgraded models should undergo thorough testing and validation to ensure stability and reliability.
2. Quickly rebuilding the model.
   Meta-learning technology is applied to the fault diagnosis model of EV powertrains. A learn-to-learn model can be trained, which has the knowledge to quickly establish new fault diagnosis models and can directly assign empirical parameters to the new model and achieve excellent performance.

Although current AI-based methods predict EV motor RUL, battery RUL and SOC, and so on are relatively accurate, there are still errors. The future of AI is aimed at reducing prediction errors. RNN models, such as LSTM, are currently the main AI technology for dealing with prediction problems, but the disadvantage of LSTM is that it cannot consider global information and cannot be calculated in parallel in the device. At present, no one has used the transformer model in EV prediction applications. It has a unique attention mechanism and can be calculated in parallel, which reduces the prediction time while improving accuracy. This will be a trend for AI to improve performance in EV applications in the future.

Furthermore, recent-emerged large-scale parametric language models such as ChatGPT have a certain impact on EV powertrain safety application. ChatGPT can provide an interactive interface for users to communicate and describe potential issues or symptoms related to the EV powertrain. It can ask clarifying questions, gather relevant information and assist users in diagnosing faults or determining the next steps for condition monitoring.

8.1 Dynamic neural network for electric powertrain health monitoring in EVs

Deep neural networks have greatly promoted the development of visual perception and NLP tasks, but most current deep models perform static reasoning graphs, that is, the network structure (depth) is fixed during the reasoning stage. This static inference mode makes the model unable to adaptively adjust the depth according to the characteristics of the input samples (such as complexity) and cannot balance the model inference efficiency and prediction accuracy. Therefore, dynamic neural networks have attracted extensive attention in recent years, and this type of neural network can automatically adjust the inference depth according to the complexity of the input samples.

The dynamic network can adaptively adjust its own structure/parameters according to the input samples in the inference stage, so it has many good characteristics that the static network cannot enjoy. In general, dynamic computation in deep learning has advantages in the following aspects:

1. Inference efficiency.
   Many dynamic networks can achieve on-demand allocation of computing resources by selectively activating modules in the model (such as network layers, convolution channels, or subnetworks), so that more easily identifiable samples It saves redundant calculations and improves reasoning efficiency in areas with less information (time or space).
2. Expression ability.
   By dynamically adjusting the structure/parameters of the network, the dynamic model can have a larger parameter space and stronger expressive ability. For example, by performing feature-based dynamic integration of multiple sets of convolutional weights, the model capacity can be significantly improved with minimal computational overhead. It is worth mentioning that commonly used attention mechanisms can also be unified into the framework of dynamic networks, since different channels, spatial regions, and temporal locations of features are dynamically redistributed during the testing phase weighted.
3. Adaptive.
   Compared with a static network with a fixed amount of computation, a dynamic network can achieve a dynamic balance of model accuracy versus efficiency in the face of changing environments (such as different hardware platforms).

4. Compatibility.
   The dynamic network is not "starting from scratch" but is compatible with other advanced technologies in the field of deep learning, so that these advanced technologies can be used to achieve SOTA performance. For example, dynamic networks can be built directly based on lightweight models or can be designed using NAS techniques. In addition, dynamic networks can also use acceleration techniques designed for traditional static models (pruning, quantization, etc.) to further improve computing efficiency.
5. Versatility.
   Many dynamic networks can be used in multiple tasks, including image classification, object detection, semantic segmentation, and so on. Additionally, techniques developed in many vision tasks can be extended to the NLP domain and vice versa.
6. Interpretability.
   Finally, some studies have pointed out the dynamic mechanism of information processing in the biological brain. Research on dynamic networks may help people build a bridge between deep models and the inner working mechanism of the brain. With dynamic networks, one can analyze: (1) Which modules of the network are activated when processing a sample; (2) which parts of the input are decisive for the final prediction.

In fact, the core idea of dynamic networks—adaptive reasoning has been studied by some researchers before the popularity of deep networks today. The most classic of these is to construct a dynamic ensemble of multiple models (in a cascaded or parallel manner) and adaptively activate these models based on input samples.

8.1.1 Sample-adaptive dynamic network

Sample-adaptive dynamic networks perform adaptive computations for each input sample. According to the subject of network dynamic change, it can be divided into two categories: dynamic structure and dynamic parameters. The former is mainly to allocate less computing resources to improve computing efficiency when processing "simple" samples; while the latter is mainly to improve the model expression ability while increasing the amount of calculation as little as possible.

8.1.2 Dynamic structure

Structural dynamics have two dimensions, one is depth and the other is width (Figure 8.4). Dynamics in depth, that is, changes in the number of network layers. Since almost all networks are stacked by multiple network layers, a more natural way to realize a dynamic structure is to selectively execute different network layers for different samples. There are two types of ideas: early exit mechanism and layer jump mechanism. The so-called early exit mechanism is to set outlets in the middle layer of the model, and according to the output of each sample at these intermediate outlets, adaptively decide whether to exit the reasoning process of the sample in advance. The early exit mechanism is equivalent to skipping the operation of all layers after a certain classifier.

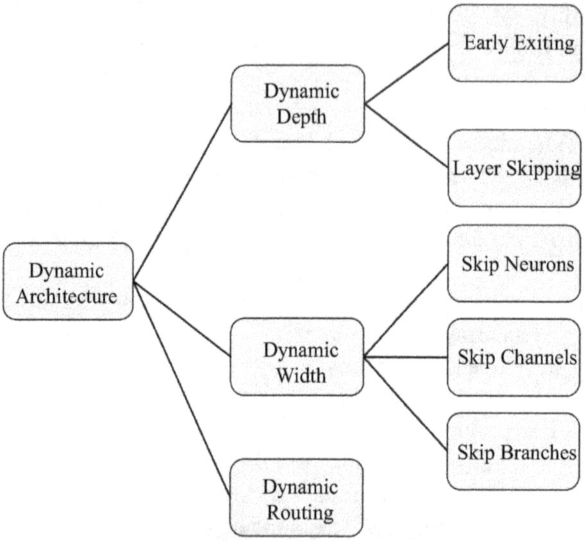

Figure 8.4 Diagram of dynamic structure

8.1.3 Two basic implementation ideas of the "early exiting" mechanism

The layer-hopping mechanism is relatively flexible. It will adaptively determine whether to execute each intermediate layer of the network for each input sample.

8.1.4 Several realization methods of dynamic layer skipping

The dynamics in the dimension of network "width" can be divided into dynamic activation of neurons, dynamic channel pruning and multi-expert hybrid system (i.e., MoE) from different scales.

The dynamic activation of neurons is easy to understand. As shown in Figure 8.5, the activation of neurons in the linear layer can usually be controlled by means of low-rank decomposition. Dynamic channel pruning in CNN, compared with the static pruning method that permanently removes some "unimportant" channels, the former can adaptively activate different convolution channels according to the sample, thus maintaining model capacity to improve computational efficiency. The multi-expert hybrid system establishes multiple "experts" through a parallel structure. These experts can be complete models or network modules, and then dynamically weight the output results of these "experts" to obtain the final prediction result.

In addition to dynamic adaptation in the breadth and depth of the network, another type of work is to first establish a super network with multiple forward channels, and then use a certain strategy to dynamically route different input samples (Figure 8.6).

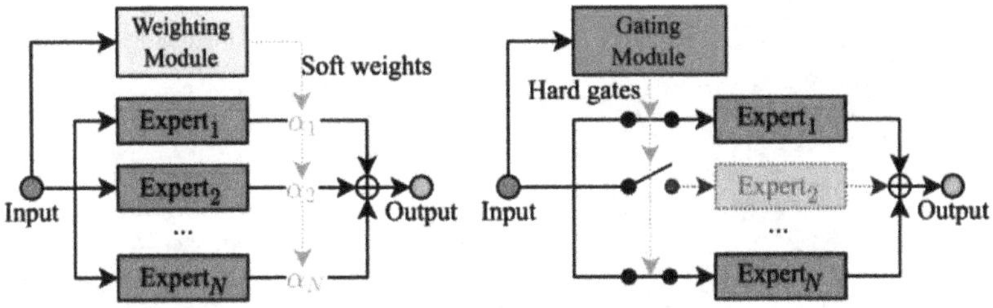

(a) Soft weights for adaptive fusion. (b) Selective execution of MoE branches.

Figure 8.5 MoE structure

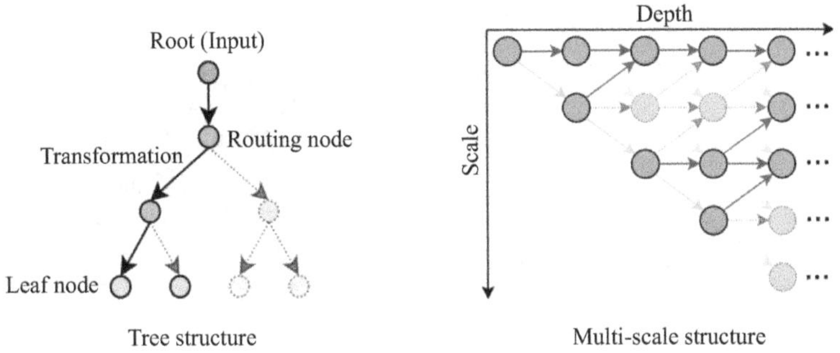

Figure 8.6 Dynamic routing in hypernetworks

8.1.5 Dynamic parameters

Networks with dynamic structures often require special structural design, training strategies or hyper-parameter adjustments (Figure 8.7). Another type of work keeps the network structure unchanged during the inference process and adaptively adjusts (some) parameters of the model according to the input samples, thereby improving the expressive ability of the model. Overall, the research on dynamic parameters can be divided into three categories: dynamic adjustment of parameters, parameter prediction, and attention-based dynamic features.

There are two core ideas for parameter dynamic adjustment: (1) For the parameters of the backbone network, use attention to reweight during the test phase, thereby improving the expressive ability of the network; (2) adaptively adjust the shape of the convolution kernel, so that the network has dynamic receptive field (Figures 8.8 and 8.9).

8.1.6 Parameter prediction

Parameter prediction is more straightforward than dynamic tuning, which predicts (part of) the parameters of the generative network directly from the input. The main

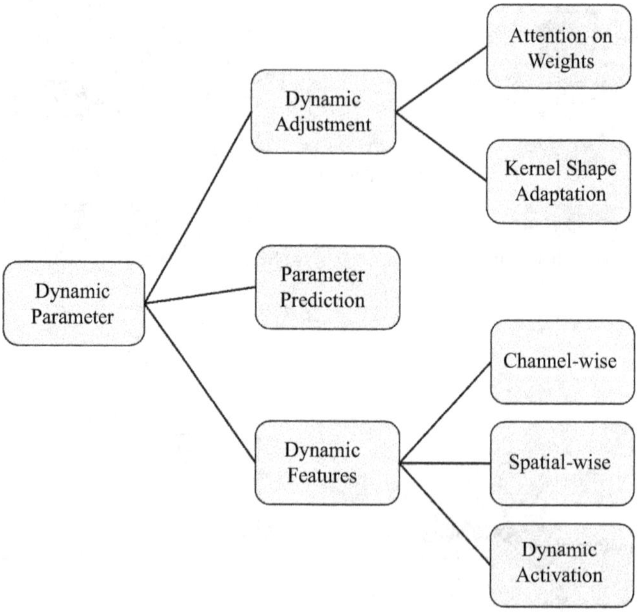

Figure 8.7 Diagram of dynamic parameters

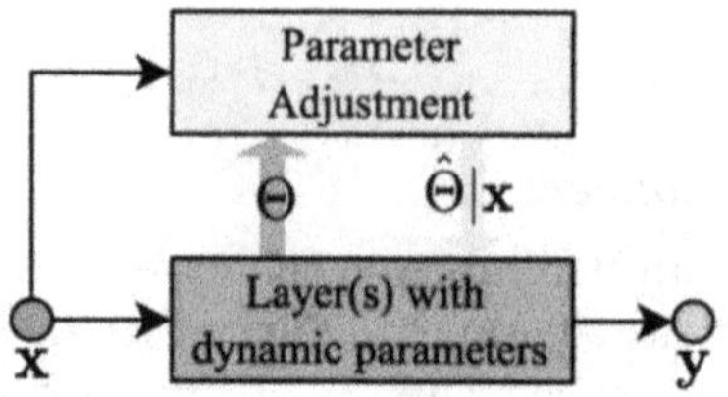

Figure 8.8 Dynamic weight adjustment of parameters

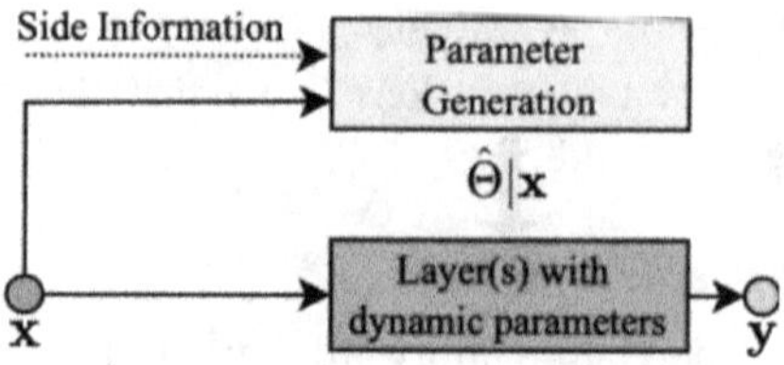

Figure 9.9 Dynamic weight prediction of parameters

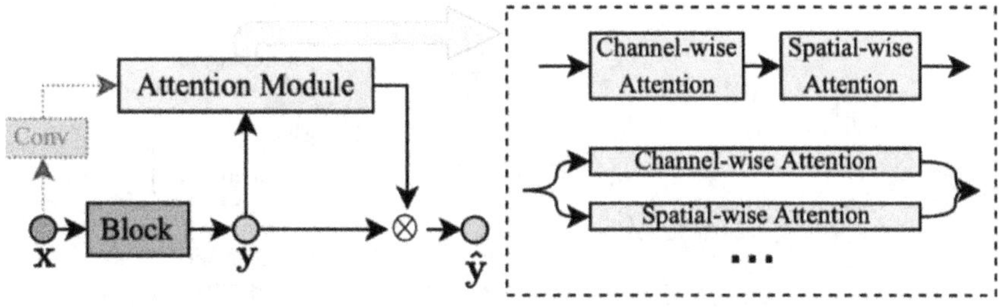

Figure 8.10 Attention-based dynamic features

effect of the dynamic parameter method is to generate more dynamic and diverse features, thereby improving the expressive ability of the model. To achieve this, an equivalent solution is to dynamically weight the features directly.

Besides channels, attention can also be used to adaptively weight different spatial locations of features (Figure 8.10). These two attention paradigms (channel and space) can be combined in various forms to further improve the model expressiveness. The previous dynamic feature methods usually reweight the features before a nonlinear activation function such as ReLU. Recently, there are also some works that directly design dynamic activation functions and replace the static activation functions in traditional networks, which can also greatly improve the expressive ability of the model.

8.1.7 Spatial adaptive dynamic network

In vision tasks, it has been shown that different spatial locations in the input play different roles in the final prediction of CNNs (Figure 8.11). To make an accurate prediction, it may only need to adaptively process a part of the spatial position in the input, without performing the same amount of calculation on different positions of the entire input image. Other studies have shown that using only lower resolutions of the input images has enabled the network to achieve good accuracy. Therefore, the practice of using the same resolution representation for all input images in traditional CNN will bring unavoidable redundant calculations. To this end, a spatially adaptive dynamic network can be designed to perform adaptive inference on the image input from a spatial perspective. According to the granularity of dynamic operations, the spatial adaptive network can be divided into three levels: pixel level, region level, and resolution level. This kind of dynamic neural network is not suitable for EV safety improvement engineering tasks, so it will not be explained too much here.

8.1.8 Time-adaptive dynamic network

Time series data (such as text, video, etc.) has great redundancy in the time dimension (Figure 8.12). Therefore, designing a dynamic network to perform adaptive calculations for data at different time locations will make the network itself more efficient.

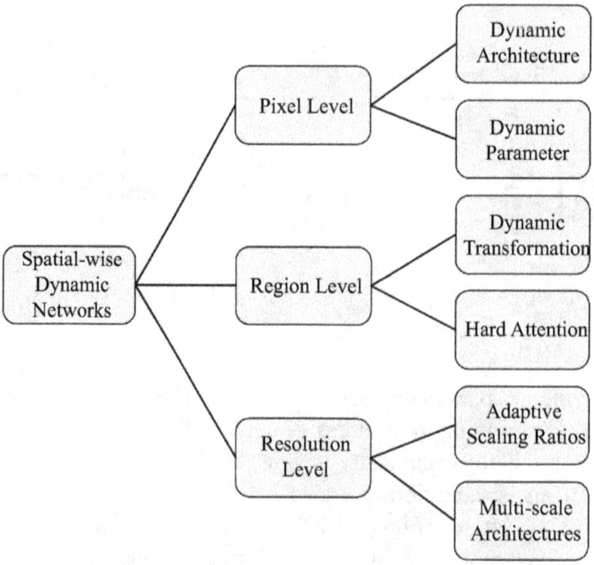

Figure 8.11 Diagram of spatial adaptive dynamic network

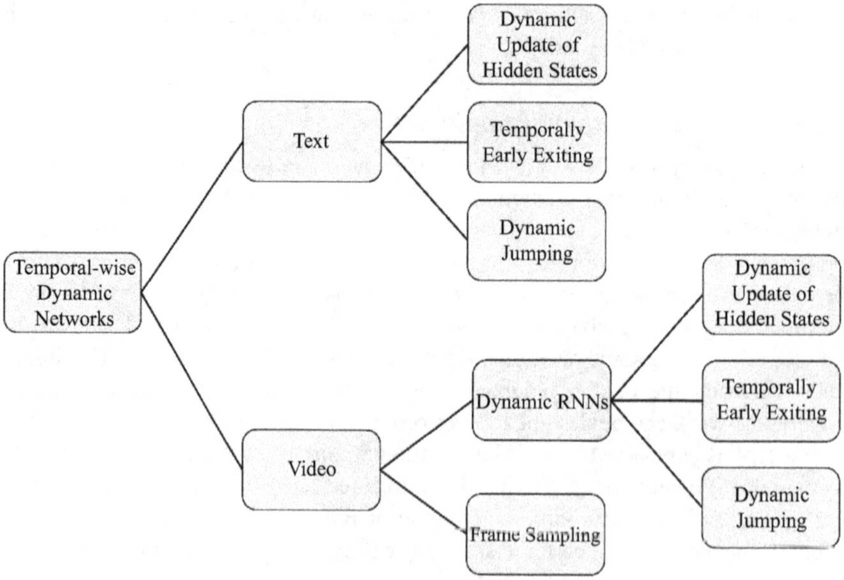

Figure 8.12 Time-adaptive dynamic network

In general, time-adaptive dynamic networks can reduce redundant calculations in two ways: (1) For some "unimportant" position inputs, assign less calculations; (2) execute only a part of the sampled time positions calculate.

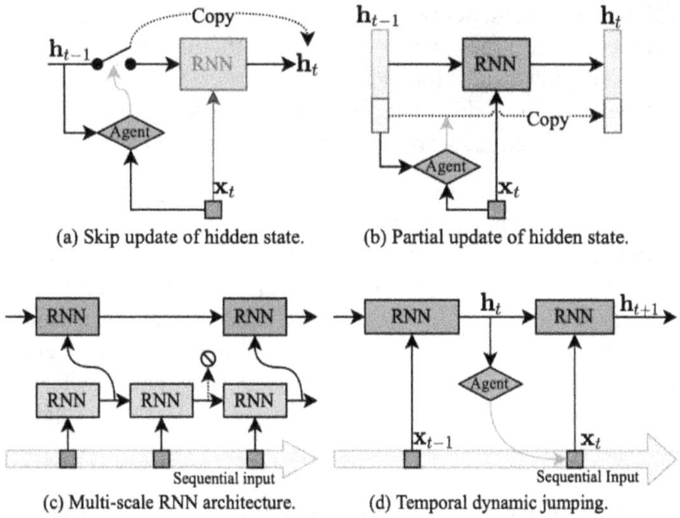

Figure 8.13 Several inference modes of time-adaptive dynamic networks [333]

Conventional static RNN updates its hidden state iteratively for the time-series data-processing flow of length *T*. However, due to the different contributions (importance) of the input to the task at different moments, using the same complexity operation for the input at each moment will lead to unavoidable redundant calculations. Therefore, different forms of dynamic RNNs can be designed to adaptively decide whether to allocate calculations or to use complex calculations according to the input at different times (Figure 8.13).

The first way is the dynamic update of the hidden state. Considering that the importance of input data is different at different times, the hidden state can be updated by adaptive calculation at each time step of the RNN, for example, the update of the hidden state can be skipped, or a rough update can be performed, or multiple the scale schema is selectively updated.

However, this method still needs to input samples at all times and then perform adaptive calculation of the required decisions. In many scenarios, only the beginning of the time series data can solve the required problem. For example, humans can have a certain understanding of the general idea of a chapter just by reading the abstract of the chapter. Therefore, an "early leave" mechanism can be adopted to terminate "reading" early at a certain intermediate moment.

In the "early exit" mechanism, the network can only decide whether to terminate the operation, but it cannot decide which positions of the data to "see" in the input. If the "jumping" mechanism mentioned previously can be used, a certain number of inputs can be skipped.

8.1.9 Open questions

The advantages of dynamic neural networks are obvious. Especially at the moment, the market urgently needs to deploy deep learning models to various mobile terminals. How to use a smaller model, consume less computation, and ensure high enough accuracy, these urgent needs are driving the research of dynamic neural networks to make great strides forward.

But despite this, this field is still in its infancy, and there are still many open issues and research directions worth exploring. We believe that there are six major issues:

1. Dynamic network structure design.
   At present, most work on network structure is dedicated to designing static networks, and most dynamic network design methods are also based on existing classic static model architectures and selectively execute different units in them. This may only be a suboptimal solution for the structural design of dynamic networks. Therefore, the dynamic network structure design for adaptive computing can further improve its performance and computing efficiency.
2. Applicability to more diverse tasks.
   At present, most dynamic networks are still only designed for classification tasks, and it is difficult to directly apply them to other vision tasks, such as object detection and semantic segmentation. The difficulty is that there is no simple criterion for judging the difficulty of samples in these tasks, and a sample contains multiple targets/pixels of different complexity. Some existing methods, such as spatial adaptive networks, can already be used for more tasks than classification. However, it is still challenging to design a unified, concise and elegant dynamic network structure that can directly serve as the backbone network for other tasks.
3. The gap between actual efficiency and theory.
   Most of the existing deep learning computing hardware and software libraries are developed for static models and are not friendly enough for the acceleration of dynamic models. As a result, the actual acceleration of dynamic models may lag behind the theoretical effect. Therefore, designing hardware-friendly dynamic networks is a valuable and challenging topic. Another interesting direction is to further optimize the computing hardware and software platform, so as to better harvest the theoretical efficiency improvement brought by the dynamic network.
4. Robustness.
   Recent work has shown that dynamic models can bring new research perspectives to the robustness of deep networks. In addition, the usual adversarial attacks focus on reducing the accuracy of the model, while for dynamic networks, the accuracy and efficiency of the model can be attacked at the same time. Therefore, the robustness of dynamic networks is an interesting and understudied topic.
5. Interpretability.
   Dynamic networks inherit the "black box" characteristics of deep neural networks, bringing research directions to explain their working mechanism. What is special is that the adaptive reasoning mechanism of dynamic networks, such

as spatial/temporal adaptability, matches the human visual system. In addition, for a given sample, it is convenient to analyze which parts of the dynamic network need to be activated to make predictions. We expect these properties to inspire new deep learning interpretability work.

6. Dynamic network theory.
   (i) Optimal decision-making problem: Decision-making is an essential operation in the reasoning process of most dynamic networks. Existing methods (based on confidence, policy network or gate function) lack certain theoretical guarantees and are not necessarily the optimal decision-making method. Therefore, designing a decision function with theoretical guarantees is a valuable research direction in the future.
   (ii) Generalization performance: In the dynamic model, different sub-networks are activated by different test samples, resulting in a certain deviation in the data distribution faced by these sub-networks during the training and testing phases. Therefore, it would be an interesting topic for a new theory of the generalization performance of dynamic networks.

8.2 Federated learning for electric powertrain health monitoring in EVs

Federated learning is an ML technique that enables a group of organizations, or groups within the same organization, to collaboratively and iteratively train and improve a shared global ML model. When using this method, data are not shared beyond a single device or group. The participating organizations form a federation that can consist of various configurations (such as geographic regions and time zones) or different business units within the same organization.

We refer to each enterprise participating in the joint modeling as a participant and divide federated learning into three categories according to the data distribution among multiple participants: horizontal federated learning, vertical federated learning, and federated transfer learning.

8.2.1 Horizontal federated learning

The essence of horizontal federated learning is the combination of samples, which is suitable for scenarios where the participants have the same format but reach different customers, that is, when there is a lot of overlap in features and little overlap in users, such as between banks in different regions, their businesses are similar (similar features), but users are different (samples are different) (Figure 8.14).

Step1: Participants download the latest model from server A.
Step2: Each participant uses local data to train the model and uploads the encrypted gradient to server A, and server A aggregates the gradients of each user to update the model parameters.
Step3: Server A returns the updated model to each participant.
Step4: Each participant updates their respective models.

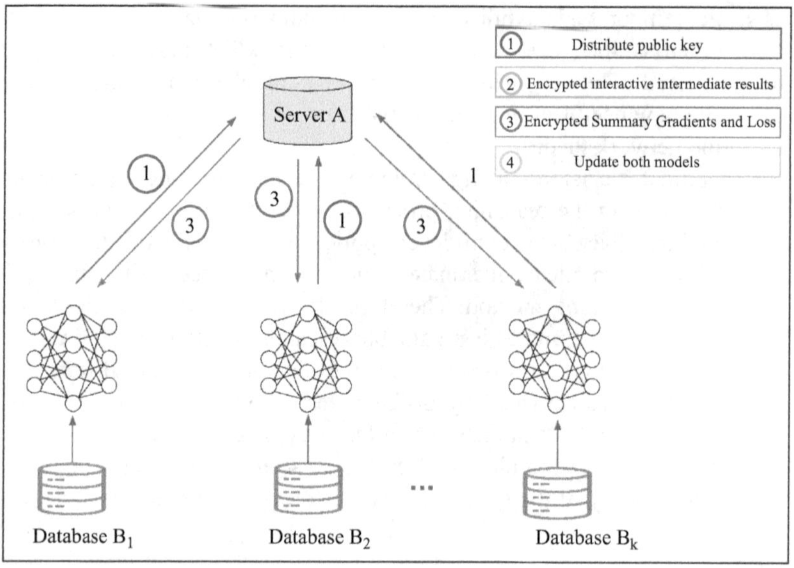

Figure 8.14 Diagram of horizontal federated learning

Interpretation of the steps: In traditional ML modeling, the data required for model training is usually collected in a data center and then the model is trained and then predicted. In horizontal federated learning, it can be regarded as sample-based distributed model training, distributing all data to different machines, each machine downloads the model from the server, then uses the local data to train the model, and then returns to the server the parameters that need to be updated. The server aggregates the returned parameters on each machine, updates the model, and feeds back the latest model to each machine.

In this process, each machine has the same and complete model, and the machines do not communicate or depend on each other. When predicting, each machine can also predict independently. This process can be regarded as a sample-based distributed model training. Google initially adopted a horizontal federation approach to solve the problem of Android phone end users updating models locally.

A sample code is given here, just to demonstrate the basic process of horizontal federated learning. The code loads the test dataset and creates the corresponding data loader *test_loader*. Then, the *test()* function is called to test the federated learning model and output the test results. The function *test()* calculates the loss and accuracy of the model on the test dataset. In practical applications, you need to make appropriate modifications and adjustments according to the data and needs, such as considering privacy protection in federated learning, optimizing model aggregation strategies, and so on.

8.2.2 Vertical federated learning

The essence of vertical federated learning is the combination of features, which is suitable for scenarios with a lot of user overlap and little feature overlap, such as supermarkets and banks in the same area. The users they reach are all residents of the area (the same sample), but the business is different (characteristics are different).

The essence of vertical federated learning is the combination of features of cross-users in different formats, such as supermarket A and bank B. In the traditional ML modeling process, two parts of data need to be concentrated in one data center, and then each user's features are joined into a piece of data to train the model, so both parties need to have user intersection (modeling based on the join result), and one side has a label. The learning steps are shown in Figure 8.15, which are divided into two steps:

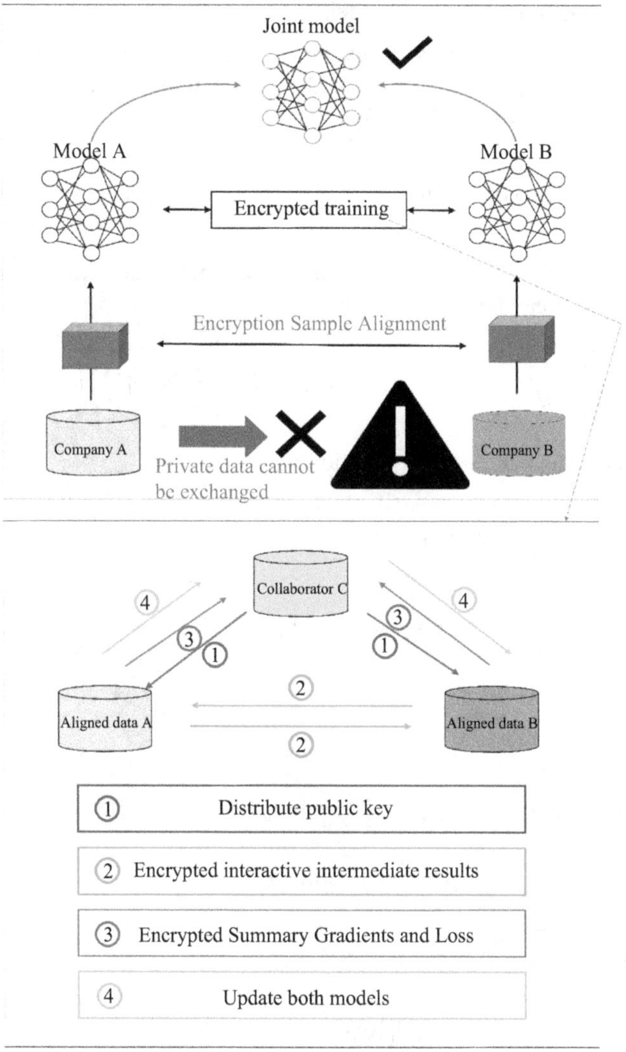

Figure 8.15 Diagram of vertical federated learning

Step 1: Encryption sample alignment. Do this at the system level, so non-cross-users are not exposed at the enterprise-aware level.
Step 2: Align samples for model encryption training: The third party C sends the public key to A and B to encrypt the data to be transmitted.
Step 3: A and B, respectively, calculate the intermediate results of the features related to themselves and encrypt and interact with each other to obtain their respective gradients and losses.
Step 4: A and B, respectively, calculate their encrypted gradients and add masks to send to C, while B calculates the encrypted loss and sends to C.
Step 5: C decrypts the gradient and loss and sends it back to A and B, A and B remove the mask and update the model.

We take linear regression as an example to illustrate its training process. There are datasets $\{x_i^A\}, i \in D_A$, $\{x_i^B, y_i^B\}, i \in D_A$, A and B initialize model parameters, respectively, Θ^A, Θ^B. Its objective function is

$$\min_{\Theta^A,\Theta^B} \sum_i \|\Theta_A x_i^A + \Theta_B x_i^B - y_i\|^2 + \frac{\lambda}{2}\left(\|\Theta_A\|^2 + \|\Theta_B\|^2\right) \tag{8.1}$$

Let $u_i^A = \Theta_A x_i^A$, $u_i^B = \Theta_B x_i^B$, and homomorphic encryption of the original objective function can be expressed as

$$[[L]] = \left[\left[\sum_i \left(\left(u_i^A + u_i^B - y_i\right)\right)^2 + \frac{\lambda}{2}\left(\|\Theta_A\|^2 + \|\Theta_B\|^2\right)\right]\right] \tag{8.2}$$

where $[[\cdot]]$ means synchronous encryption and

$$[[L_A]] = \left[\left[\sum_i \left(\left(u_i^A\right)\right)^2 + \frac{\lambda}{2}\|\Theta_A\|^2\right]\right], \tag{8.3}$$

$$[[L_B]] = \left[\left[\sum_i \left(\left(u_i^B - y_i\right)\right)^2 + \frac{\lambda}{2}{\Theta_B}^2\right]\right], \tag{8.4}$$

$$[[L_{AB}]] = 2\sum_i \left(\left[\left[u_i^A\right]\right]\left(u_i^B - y_i\right)\right) \tag{8.5}$$

So there is $[[L]] = [[L_A]] + [[L_B]] + [[L_{AB}]]$, similarly $[[d_i]] = \left[\left[u_i^A\right]\right] + \left[\left[u_i^B - y_i\right]\right]$.

The gradient can be expressed as follows:

$$\left[\left[\frac{\partial L}{\partial \Theta_A}\right]\right] = \sum_i [[d_i]] x_i^A + [[\lambda \Theta_A]] \tag{8.6}$$

$$\left[\left[\frac{\partial L}{\partial \Theta_B}\right]\right] = \sum_i [[d_i]] x_i^B + [[\lambda \Theta_B]] \tag{8.7}$$

The specific training steps are shown in Table 8.2.

Table 8.2 Specific training steps of vertical federated learning

Step	Party A	Party B	Party C
Step 1	Initialize Θ_A	Initialize Θ_B	Create an encryption key pair send public key to A and B
Step 2	Compute $\left[\left[u_i^A\right]\right]$, $[[L_A]]$ and send to B	Compute $\left[\left[u_i^B\right]\right]$,$\left[\left[d_i^B\right]\right]$,$[[L]]$, send $\left[\left[d_i^B\right]\right]$ to A, send $[[L]]$ to C	
Step 3	Initialize R_A, compute $\left[\left[\frac{\partial L}{\partial \Theta_A}\right]\right] + [[R_A]]$, and send to C	Initialize R_B, compute $\left[\left[\frac{\partial L}{\partial \Theta_B}\right]\right] + [[R_B]]$, and send to C	C decrypt L, send $\frac{\partial L}{\partial \Theta_A} + R_A$ to A, $\frac{\partial L}{\partial \Theta_B} + R_B$ to B
Step 4	Update Θ_A	Update Θ_B	–
What is obtained	Θ_A	Θ_B	–

During the whole process, the participants do not know the data and characteristics of the other party, and after the training, the participants only get the model parameters of their own side, that is, the half model.

Prediction process: Since each participant can only obtain the model parameters related to itself, the prediction requires the cooperation of both parties, as shown in Figure 8.16. The result of joint modeling is that both parties can obtain data protection, jointly improve the model effect, and the model has no loss.

The provided example code demonstrates an example of vertical federated learning. In this example, we assume that there are two participants (Participants 1 and 2), each with different characteristics and labels. We use two data loaders *train_loader_1* and *train_loader_2* to load the data of the two parties for training, respectively. During training, we create the global model and the optimizer, using the cross-entropy loss function *nn.BCE*.

8.2.3 Federated transfer learning

Federated transfer learning can be considered when there is little overlap in features and samples between participants, such as the union between banks and supermarkets in different regions. It is mainly applicable to scenarios based on deep neural networks. The steps of federated transfer learning are similar to vertical federated learning, except that the intermediate transfer results are different (in fact, the intermediate transfer results of each model are different). Here, we will focus on the idea of federal migration (Figure 8.17).

Source domain dataset is $D_S = \left\{\left(x_i^S, y_i^S\right)\right\}_{i=1}^{N_S}$ and target domain dataset is $D_T = \left\{\left(x_j^T\right)\right\}_{j=1}^{N_T}$. We assume that there is a common sample $D_{ST} = \left\{\left(x_i^S, x_i^T\right)\right\}_{i=1}^{N_{ST}}$ between the source domain and the target domain. For its common samples $D_C = \left\{\left(x_i^T, y_i^S\right)\right\}_{i=1}^{N_C}$, u_S, u_T are the hidden layer feature invariants between the

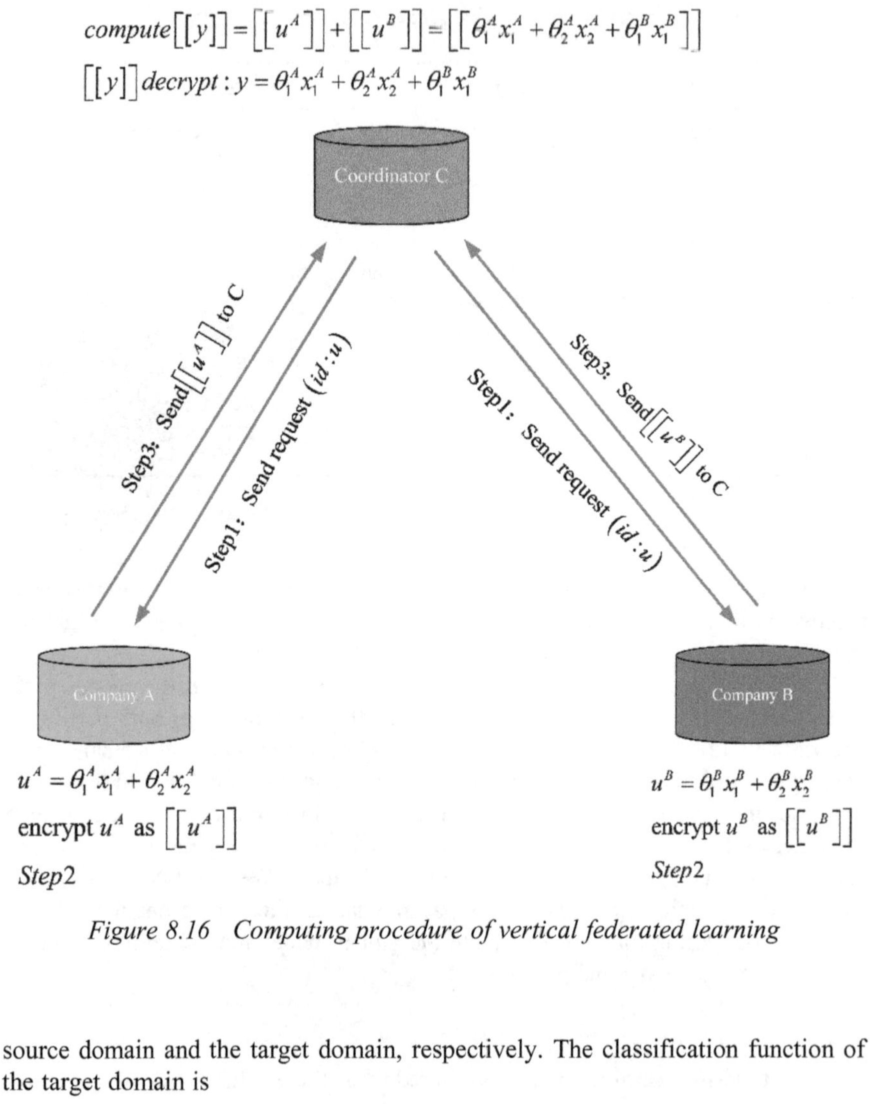

Figure 8.16 Computing procedure of vertical federated learning

source domain and the target domain, respectively. The classification function of the target domain is

$$\varphi\left(u_j^T\right) = \frac{1}{N}\sum_i^{N_S} y_i^S u_i^S \left(u_j^T\right)' = \Phi^S \Omega\left(u_j^T\right) \tag{8.8}$$

The objective function can be defined as

$$\operatorname{argmin}_{\Theta^S,\Theta^T} L_1 = \sum_i^{N_C} l_1\left(y_i^S, \varphi\left(u_i^T\right)\right) \tag{8.9}$$

$$\operatorname{argmin}_{\Theta^S,\Theta^T} L_2 = \sum_i^{N_{ST}} l_1\left(u_i^S, u_i^T\right) \tag{8.10}$$

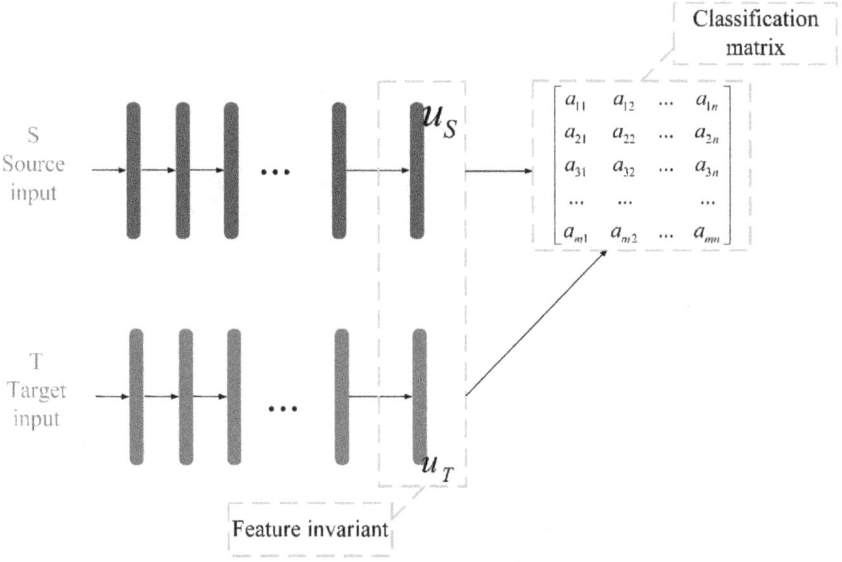

Figure 8.17 Federated transfer learning

The overall objective function is

$$\text{argmin}_{\Theta^S,\Theta^T} L = L_1 + \lambda L_2 + \frac{\lambda}{2}\left(\|\Theta^S\|^2 + \|\Theta^T\|^2\right) \tag{8.11}$$

Using the BP algorithm, according to the objective function L, the gradients of Θ^S, Θ^T are, respectively, calculated, and the two parties interactively calculate the intermediate results needed for the gradient and loss, and iterate until convergence. The whole learning process is to use the common samples between S and T to learn the respective feature invariant representations u_S, u_T, and at the same time use all the samples label y_S and the invariant feature u_S of S to learn the classifier. In this stage, "federalization" is reflected in S, and T can jointly learn a model through the intermediate results of safe interaction, and "transfer" is reflected in the fact that T migrates S's classification ability. When predicting, u_T depends on the classifier composed of u_S, y_S, so it needs to cooperate with the vertical federation to complete.

A sample code is provided here, and more adjustments and modifications may be required in actual applications, such as considering batch normalization, data enhancement, learning rate adjustment, and so on, to optimize the performance of the model. The code loads the test dataset and creates the corresponding data loader *test_loader*. Then, set the model into evaluation mode by using *model.eval()* and disable gradient computation using *torch.no_grad()* in order to speed up the inference process. Iterate over each batch of images and labels on the test dataset, compute the output of the model, and compare with the actual labels to calculate the accuracy. Finally, output the accuracy of the model on the test set.

Appendix

The appendix section consists of 21 subsections, each providing practical example code covering various topics and programming languages. Here are the names and brief descriptions of each subsection:

1. FFT-Python: Introduces example code for performing fast Fourier transform (FFT) using Python.
2. HT-Python: Introduces example code for performing Hadamard transform using Python.
3. STFT-Python: Introduces example code for performing short-time Fourier transform (STFT) using Python.
4. DWT-Python: Introduces example code for performing discrete wavelet transform (DWT) using Python.
5. EMD-Python: Introduces example code for performing empirical mode decomposition (EMD) using python.
6. PCA-Python: Introduces example code for performing principal component analysis (PCA) using Python.
7. SVD-Python: Introduces example code for performing singular value decomposition (SVD) using Python.
8. SGD-Python: Introduces example code for implementing stochastic gradient descent (SGD) using Python.
9. Adam-Python: Introduces example code for implementing the Adam optimization algorithm using Python.
10. Basic operations-PyTorch: Introduces example code for basic operations in PyTorch.
11. Variable in PyTorch: Introduces example code for using variables in PyTorch.
12. Activation function in torch: Introduces example code for using activation functions in PyTorch.
13. Example: Logistic regression: Introduces example code for implementing logistic regression.
14. Example: Classification model-TensorFlow/Keras: Introduces example code for implementing a classification model using TensorFlow/Keras.
15. GAN for text generation-PyTorch: Introduces example code for text generation using generative adversarial networks (GANs) in PyTorch.

16. Example of RF regression: Introduces example code for implementing random forest regression.
17. Example: RUL prediction of IGBT: Introduces example code for predicting the remaining useful life (RUL) of insulated gate bipolar transistors (IGBTs).
18. Example: IGBT open-circuit detection: Introduces example code for detecting open-circuit faults in IGBTs.
19. Example: Bearing RUL prediction: Introduces example code for predicting the Remaining Useful Life (RUL) of bearings.
20. Horizontal federated learning: Introduces example code for horizontal federated learning.
21. Vertical federated learning: Introduces example code for vertical federated learning.
22. Federated transfer learning: Introduces example code for federated transfer learning.

These example codes will help readers quickly understand and practice various signal processing, machine learning, deep learning techniques, as well as applications of federated learning. Each subsection provides example code that can serve as a starting point for readers to further learn and apply these techniques.

1 FFT-Python

```
import numpy as np

import matplotlib.pyplot as plt


# Generate input signal

t = np.linspace(0, 1, 1000)  # Time vectors

f = 10  # Input signal frequency

x = np.sin(2 * np.pi * f * t)  # Input signals


# Execute FFT

X = np.fft.fft(x)


# Calculation of frequency axis
```

```
freq = np.fft.fftfreq(len(t), t[1] - t[0])

# Plotting the spectrum

plt.plot(freq, np.abs(X))
plt.xlabel('Frequency (Hz)')
plt.ylabel('Magnitude')
plt.title('FFT Spectrum')
plt.show()
```

2 HT-Python

```
import numpy as np
import matplotlib.pyplot as plt
from scipy.signal import hilbert

# Generate input signalt = np.linspace(0, 1, 1000)
frequency = 10
amplitude = np.sin(2 * np.pi * frequency * t)

# Perform the Hilbert transform

analytic_signal = hilbert(amplitude)
amplitude_envelope = np.abs(analytic_signal)

# Plot the original signal and the transformed envelope signal

plt.figure(figsize=(8, 6))
plt.subplot(2, 1, 1)
plt.plot(t, amplitude, 'b')
```

```
plt.xlabel('Time')
plt.ylabel('Amplitude')
plt.title('Original Signal')

plt.subplot(2, 1, 2)
plt.plot(t, amplitude_envelope, 'r')
plt.xlabel('Time')
plt.ylabel('Amplitude')
plt.title('Amplitude Envelope')

plt.tight_layout()
plt.show()
```

3 STFT-Python

```
import numpy as np
import matplotlib.pyplot as plt
from scipy.signal import stft

# Generate input signal
t = np.linspace(0, 1, 1000)
frequency = 10
amplitude = np.sin(2 * np.pi * frequency * t)

# Perform short-time Fourier transform

f, t, Zxx = stft(amplitude, nperseg=100, noverlap=50)

# Plotting STFT results
```

```
plt.figure(figsize=(8, 6))
plt.pcolormesh(t, f, np.abs(Zxx), shading='auto', cmap='jet')
plt.colorbar(label='Magnitude')
plt.title('STFT')
plt.xlabel('Time')
plt.ylabel('Frequency')
plt.ylim(0, 2 * frequency)  # Set y-axis frequency range
plt.tight_layout()
plt.show()
```

4 DWT-Python

```
import pywt
import numpy as np
import matplotlib.pyplot as plt

# Generate input signal

t = np.linspace(0, 1, 1000)
frequency = 10
amplitude = np.sin(2 * np.pi * frequency * t)

# Execute DWT

wavelet = 'db4'  # Wavelet types
levels = 5  # Decomposition layers
coeffs = pywt.wavedec(amplitude, wavelet, level=levels)

# Plotting DWT coefficients

plt.figure(figsize=(8, 6))
```

```
for i, coeff in enumerate(coeffs):
    plt.subplot(levels+1, 1, i+1)
    plt.plot(coeff)
    plt.ylabel(f'Level {i}')
plt.xlabel('Sample')
plt.tight_layout()
plt.show()
```

5 EMD-Python

```
import numpy as np
import matplotlib.pyplot as plt
from PyEMD import EMD

# Generate input signal
t = np.linspace(0, 1, 1000)
frequency1 = 10
amplitude1 = np.sin(2 * np.pi * frequency1 * t)
frequency2 = 20
amplitude2 = np.sin(2 * np.pi * frequency2 * t)
input_signal = amplitude1 + amplitude2

# Execute EMD

emd = EMD()
imfs = emd(input_signal)

# Plotting IMFs

plt.figure(figsize=(8, 6))
plt.subplot(len(imfs)+1, 1, 1)
```

```
plt.plot(input_signal)
plt.title('Input Signal')
plt.xlabel('Sample')
for i, imf in enumerate(imfs):
    plt.subplot(len(imfs)+1, 1, i+2)
    plt.plot(imf)
    plt.ylabel(f'IMF {i+1}')
    plt.xlabel('Sample')
plt.tight_layout()
plt.show()
```

6 PCA-Python

```
import numpy as np
from sklearn.decomposition import PCA

# Generate sample dataX = np.array([[1, 2, 3], [4, 5, 6], [7, 8, 9]])

# Create PCA object and fit it
pca = PCA(n_components=2) # Set the number of principal components to be retained
pca.fit(X)

# Convert data to principal component space

X_transformed = pca.transform(X)

# Print the results of the principal component analysis

print("Original data: ")
print(X)
print("\nResults of Principal Component Analysis : ")
```

```
print(X_transformed)

print("\nVariance contribution ratio of principal components : ")

print(pca.explained_variance_ratio_)
```

7 SVD-Python

```
import numpy as np

def svd(A):
    # Calculate the transpose matrix of A
    A_T = A.T
    # Calculate the product of A and the transpose matrix of A
    A_A_T = np.dot(A, A_T)

    # Calculate the eigenvalues and eigenvectors of A
    eigenvalues, eigenvectors = np.linalg.eig(A_A_T)

    # Sorting of eigenvalues
    sorted_indices = np.argsort(eigenvalues)[::-1]
    eigenvalues = eigenvalues[sorted_indices]
    eigenvectors = eigenvectors[:, sorted_indices]

        # Calculating singular values and right singular vectors

    singular_values = np.sqrt(eigenvalues)
    right_singular_vectors = np.dot(A_T, eigenvectors) / singular_values

    # Calculating the left singular vector
    left_singular_vectors = eigenvectors

    # Returns the result of the singular value decomposition
```

```
    return left_singular_vectors, singular_values, right_singular_vectors.T

# Generate sample data
X = np.array([[1, 2, 3], [4, 5, 6], [7, 8, 9]])

# Performing singular value decomposition
U, S, Vt = svd(X)

# Print the results of the singular value decomposition
print("Left singular matrix U : ")
print(U)
print("\nSingular value matrix S : ")
print(np.diag(S))
print("\nRight singular matrix V^T : ")
print(Vt)
```

8 SGD-Python

```
import numpy as np

def sgd(X, y, learning_rate=0.01, epochs=100, batch_size=1):
    # Get dataset size and number of features
    num_samples, num_features = X.shape
    # Initialize model parameters
    weights = np.zeros(num_features)
    bias = 0

    # Iterative training for epoch in range(epochs):
        # Randomly disrupted data sets
```

```
        permutation = np.random.permutation(num_samples)
        X = X[permutation]
        y = y[permutation]

        # Training by batch
        for i in range(0, num_samples, batch_size):
            # Get data for the current batch
            X_batch = X[i:i+batch_size]
            y_batch = y[i:i+batch_size]

            # Calculate predicted values
            y_pred = np.dot(X_batch, weights) + bias

            # Calculating gradients
            gradient_weights = np.dot(X_batch.T, (y_pred - y_batch)) / batch_size
            gradient_bias = np.mean(y_pred - y_batch)

            # Update model parameters
            weights -= learning_rate * gradient_weights
            bias -= learning_rate * gradient_bias

    return weights, bias

# Generate sample dataX = np.array([[1, 2], [2, 3], [3, 4], [4, 5]])
y = np.array([2, 3, 4, 5])

# Perform stochastic gradient descent training

weights, bias = sgd(X, y, learning_rate=0.1, epochs=100, batch_size=1)

# Print the parameters of the trained model
```

```
print("Weights:", weights)
print("Bias:", bias)
```

9 Adam-Python

```
import numpy as np

def adam_optimizer(X, y, learning_rate=0.01, epochs=100, batch_size=1, beta1=0.9,
beta2=0.999, epsilon=1e-8):
    # Get dataset size and number of features
    num_samples, num_features = X.shape
    # Initialize model parameters
    weights = np.zeros(num_features)
    bias = 0
    # Initialize momentum and second order moment terms
    m_weights = np.zeros(num_features)
    m_bias = 0
    v_weights = np.zeros(num_features)
    v_bias = 0

    # Iterative training
    for epoch in range(epochs):
        # Randomly disrupted data sets
        permutation = np.random.permutation(num_samples)
        X = X[permutation]
        y = y[permutation]

        # Training by batch
        for i in range(0, num_samples, batch_size):
```

```
            # Get data for the current batch
            X_batch = X[i:i+batch_size]
            y_batch = y[i:i+batch_size]

            # Calculate predicted values
            y_pred = np.dot(X_batch, weights) + bias

            # Calculating gradients
            gradient_weights = np.dot(X_batch.T, (y_pred - y_batch)) / batch_size
            gradient_bias = np.mean(y_pred - y_batch)

            # Update momentum and second order moment terms
            m_weights = beta1 * m_weights + (1 - beta1) * gradient_weights
            m_bias = beta1 * m_bias + (1 - beta1) * gradient_bias
            v_weights = beta2 * v_weights + (1 - beta2) * np.square(gradient_weights)
            v_bias = beta2 * v_bias + (1 - beta2) * np.square(gradient_bias)

            # Correction of deviations
            m_weights_hat = m_weights / (1 - beta1 ** (epoch * num_samples + i + 1))
            m_bias_hat = m_bias / (1 - beta1 ** (epoch * num_samples + i + 1))
            v_weights_hat = v_weights / (1 - beta2 ** (epoch * num_samples + i + 1))
            v_bias_hat = v_bias / (1 - beta2 ** (epoch * num_samples + i + 1))

            # Update model parameters
            weights -= learning_rate * m_weights_hat / (np.sqrt(v_weights_hat) + epsilon)
            bias -= learning_rate * m_bias_hat / (np.sqrt(v_bias_hat) + epsilon)

    return weights, bias

# Generate sample dataX = np.array([[1, 2], [2, 3], [3, 4], [4, 5]])
y = np.array([2, 3, 4, 5])
```

```
# Perform Adam optimization trainingweights, bias = adam_optimizer(X, y,
learning_rate=0.1, epochs=100, batch_size=1, beta1=0.9, beta2=0.999, epsilon=1e-8)

# Print the parameters of the trained modelprint("Weights:", weights)
print("Bias:", bias)
```

10 Basic operations-PyTorch

```
import torch
import numpy as np
np_data = np.arange(6).reshape((2, 3))
torch_data = torch.from_numpy(np_data)
tensor2array = torch_data.numpy()
print(
    '\nnumpy array:', np_data,
    '\ntorch tensor:', torch_data,
    '\ntensor to array:', tensor2array,
)
# Absolute value
data = [-1, -2, 1, 2]
tensor = torch.FloatTensor(data)
print(
    '\nabs',
    '\nnumpy: ', np.abs(data),
    '\ntorch: ', torch.abs(tensor)
)
# Sin
print(
    '\nsin',
    '\nnumpy: ', np.sin(data),
```

```
    '\ntorch: ', torch.sin(tensor)
)
# Mean
print(
    '\nmean',
    '\nnumpy: ', np.mean(data),
    '\ntorch: ', torch.mean(tensor)
)
# matrix multiplication
data = [[1,2], [3,4]]
tensor = torch.FloatTensor(data)  # Convert to 32-bit floating point tensor
print(
    '\nmatrix multiplication (matmul)',
    '\nnumpy: ', np.matmul(data, data),
    '\ntorch: ', torch.mm(tensor, tensor)
)
```

11 Variable in PyTorch

```
import torch
from torch.autograd import Variable # import Variable module of torch
tensor = torch.FloatTensor([[1,2],[3,4]])
variable = Variable(tensor, requires_grad=True)
print(tensor)

# Variable calculation, gradient
t_out = torch.mean(tensor*tensor)
v_out = torch.mean(variable*variable)
print(t_out)
print(v_out)
```

12 Activation function in torch

```
import torch
import torch.nn.functional as F    # The activation functions are all here
from torch.autograd import Variable

# Make some fake data to watch the image
x = torch.linspace(-5, 5, 200)    #example data
x = Variable(x)
x_np = x.data.numpy()
# Several commonly used activation functions
y_relu = F.relu(x).data.numpy()
y_sigmoid = F.sigmoid(x).data.numpy()
y_tanh = F.tanh(x).data.numpy()
y_softplus = F.softplus(x).data.numpy()
```

13 Example: logistic regression

```
#Create some fake data to simulate the real situation
import torch
import matplotlib.pyplot as plt
x = torch.unsqueeze(torch.linspace(-1, 1, 100), dim=1)  # x data
y = x.pow(2) + 0.2*torch.rand(x.size())  # noisy y data
# Build a neural network
import torch.nn.functional as F      # activation function
class Net(torch.nn.Module):
    def __init__(self, n_feature, n_hidden, n_output):
        super(Net, self).__init__()
        # Define what form each layer uses
        self.hidden = torch.nn.Linear(n_feature, n_hidden)  # Hidden layer linear output
        self.predict = torch.nn.Linear(n_hidden, n_output)  # Output layer linear output
```

```
    def forward(self, x):  #
        # Forward propagation of the input value, the neural network analyzes the
output value
x = F.relu(self.hidden(x))     # activation function (linear value of hidden layer)
        x = self.predict(x)          # output value
        return x
net = Net(n_feature=1, n_hidden=10, n_output=1)
# Training network
optimizer = torch.optim.SGD(net.parameters(), lr=0.2) #All parameters and learning
rate passed to net
loss_func = torch.nn.MSELoss()     # The formula for calculating the error between
the predicted value and the actual value (mean square error)
for t in range(100):
    prediction = net(x)     # Feed the net training data x, output the predicted value
    loss = loss_func(prediction, y)     # Calculate the error of both
    optimizer.zero_grad()   # Clear the residual update parameter value from the
previous step
loss.backward()        # Error backpropagation, calculate parameter update value
optimizer.step()       # Applies parameter update values to net's parameters
```

14 Example: Classification model-TensorFlow/Keras

```
# Build a machine learning model
model = tf.keras.models.Sequential([
  tf.keras.layers.Flatten(input_shape=(28, 28)),
  tf.keras.layers.Dense(128, activation='relu'),
  tf.keras.layers.Dropout(0.2),
  tf.keras.layers.Dense(10)
])
predictions = model(x_train[:1]).numpy()
# logits converted to probabilities for each class
tf.nn.softmax(predictions).numpy()
# efine the training loss function
```

```
loss_fn = tf.keras.losses.SparseCategoricalCrossentropy(from_logits=True)
# Compile model
model.compile(optimizer='adam',
           loss=loss_fn,
           metrics=['accuracy'])

 # Train and evaluate model
 model.fit(x_train, y_train, epochs=5)
 model.evaluate(x_test,  y_test, verbose=2)
# Packaging the trained model
probability_model = tf.keras.Sequential([
 model,
 tf.keras.layers.Softmax()
])
```

15 GAN for text generation-PyTorch

```
import torch
import torch.nn as nn
import torch.optim as optim
from torch.utils.data import DataLoader
from torch.utils.tensorboard import SummaryWriter

# Define the generator model
class Generator(nn.Module):
    def __init__(self, input_dim, hidden_dim, output_dim):
        super(Generator, self).__init__()
        self.hidden_dim = hidden_dim
        self.embedding = nn.Embedding(input_dim, hidden_dim)
        self.rnn = nn.GRU(hidden_dim, hidden_dim, batch_first=True)
        self.fc = nn.Linear(hidden_dim, output_dim)
```

```
    def forward(self, x):
        embedded = self.embedding(x)
        output, _ = self.rnn(embedded)
        output = self.fc(output)
        return output

# Define the discriminator model
class Discriminator(nn.Module):
    def __init__(self, input_dim, hidden_dim):
        super(Discriminator, self).__init__()
        self.hidden_dim = hidden_dim
        self.embedding = nn.Embedding(input_dim, hidden_dim)
        self.rnn = nn.GRU(hidden_dim, hidden_dim, batch_first=True)
        self.fc = nn.Linear(hidden_dim, 1)
        self.sigmoid = nn.Sigmoid()

    def forward(self, x):
        embedded = self.embedding(x)
        output, _ = self.rnn(embedded)
        output = self.fc(output)
        output = self.sigmoid(output)
        return output

# Define the training function
def train(generator, discriminator, dataloader, num_epochs, device):
    generator.to(device)
    discriminator.to(device)

    criterion = nn.BCELoss()
    gen_optimizer = optim.Adam(generator.parameters(), lr=0.001)
```

```
dis_optimizer = optim.Adam(discriminator.parameters(), lr=0.001)

for epoch in range(num_epochs):
    for real_data in dataloader:
        real_data = real_data.to(device)
        batch_size = real_data.size(0)


        dis_optimizer.zero_grad()
        real_labels = torch.ones(batch_size, 1).to(device)
        fake_labels = torch.zeros(batch_size, 1).to(device)

        real_output = discriminator(real_data)
        real_loss = criterion(real_output, real_labels)

        z = torch.randint(input_dim, size=(batch_size, seq_length)).to(device)
        fake_data = generator(z).detach()
        fake_output = discriminator(fake_data)
        fake_loss = criterion(fake_output, fake_labels)

        dis_loss = real_loss + fake_loss
        dis_loss.backward()
        dis_optimizer.step()

        gen_optimizer.zero_grad()
        fake_output = discriminator(fake_data)
        gen_loss = criterion(fake_output, real_labels)

        gen_loss.backward()
        gen_optimizer.step()
```

```
    print(f"Epoch [{epoch+1}/{num_epochs}], Discriminator Loss: {dis_loss.item()},
Generator Loss: {gen_loss.item()}")

# Data preprocessing and loading

# Create generator and discriminator instances
input_dim = len(vocab)
hidden_dim = 100
output_dim = len(vocab)
generator = Generator(input_dim, hidden_dim, output_dim)
discriminator = Discriminator(input_dim, hidden_dim)

# Train the GAN model
train(generator, discriminator, dataloader, num_epochs, device)
```

16 Example of RF regression

```
from sklearn.ensemble import RandomForestRegressor
from sklearn.metrics import mean_squared_error
from sklearn.model_selection import train_test_split
from sklearn.datasets import load_boston

# Load the Boston housing dataset
boston = load_boston()
X = boston.data
y = boston.target

# Split the data into training and testing sets
X_train, X_test, y_train, y_test = train_test_split(X, y, test_size=0.2, random_state=42)

# Create a Random Forest Regression model
```

```
rf = RandomForestRegressor(n_estimators=100, random_state=42)

# Train the model
rf.fit(X_train, y_train)

# Make predictions on the test set
y_pred = rf.predict(X_test)

# Evaluate the model
mse = mean_squared_error(y_test, y_pred)
print("Mean Squared Error:", mse)
```

17 Example: RUL prediction of IGBT

```
import numpy as np
import torch
import torch.nn as nn
import pandas as pd
import torch.optim as optim
from torch.utils.data import Dataset, DataLoader
from sklearn.preprocessing import MinMaxScaler

# Define the Transformer model
class TransformerModel(nn.Module):
    def __init__(self, input_dim, output_dim, num_heads, num_layers):
        super(TransformerModel, self).__init__()
        self.transformer = nn.Transformer(d_model=input_dim,
nhead=num_heads,
num_encoder_layers=num_layers)
        self.fc = nn.Linear(input_dim, output_dim)
```

```
    def forward(self, x):
        x = self.transformer(x, x)
        x = self.fc(x)
        return x.squeeze(1)

# Define a custom dataset for IGBT RUL prediction
class IGBTDataset(Dataset):
    def __init__(self, data):
        self.data = data

    def __len__(self):
        return len(self.data)

    def __getitem__(self, idx):
        return self.data[idx]

# Sample data
time = pd.read_csv(time.CSV) # Time steps
current = pd.read_csv(current.CSV)  # IGBT current
rul = pd.read_csv(rul.CSV)  # Remaining useful life

# Normalize the data
scaler = MinMaxScaler()
current = scaler.fit_transform(current.reshape(-1, 1))
rul = scaler.fit_transform(rul.reshape(-1, 1))

# Prepare the data for training
data = np.concatenate((current, rul), axis=1)
dataset = IGBTDataset(data)
dataloader = DataLoader(dataset, batch_size=1, shuffle=True)
```

```
# Define the model hyperparameters
input_dim = 1  # Dimension of input sequence (IGBT current)
output_dim = 1  # Dimension of output (RUL)
num_heads = 2  # Number of attention heads in the Transformer model
num_layers = 2  # Number of encoder layers in the Transformer model

# Create the Transformer model
model = TransformerModel(input_dim, output_dim, num_heads, num_layers)

# Define the loss function and optimizer
criterion = nn.MSELoss()
optimizer = optim.Adam(model.parameters(), lr=0.001)

# Training loop
num_epochs = 100
for epoch in range(num_epochs):
    for batch in dataloader:
        inputs = batch[:, :-1]
        targets = batch[:, -1]

        optimizer.zero_grad()
        outputs = model(inputs.unsqueeze(1))

        loss = criterion(outputs, targets)
        loss.backward()
        optimizer.step()

    if (epoch + 1) % 10 == 0:
        print(f"Epoch {epoch+1}/{num_epochs}, Loss: {loss.item()}")

# Predict RUL for a new IGBT current value
```

```
new_current = np.array([30]).reshape(-1, 1)
new_current = scaler.transform(new_current)
inputs = torch.tensor(new_current).float()

predicted_rul = model(inputs.unsqueeze(0))
predicted_rul = scaler.inverse_transform(predicted_rul.detach().numpy())

print("Predicted RUL:", predicted_rul[0][0])
```

18 Example: IGBT open-circuit detection

```
import numpy as np
from sklearn.svm import SVC
from sklearn.metrics import accuracy_score
from sklearn.model_selection import train_test_split
from scipy.stats import entropy, kurtosis, skew

# Sample data
normal_data = np.random.normal(loc=0, scale=1, size=(1000, 100))  # Normal
operation data
faulty_data = np.random.normal(loc=2, scale=1, size=(200, 100))  # IGBT open-circuit
fault data

# Function to extract features from current signals
def extract_features(data):
    features = []
    for signal in data:
        # Calculate statistical features
        mean = np.mean(signal)
        std = np.std(signal)
        max_value = np.max(signal)
        min_value = np.min(signal)
        entropy_value = entropy(signal)
```

```
    kurtosis_value = kurtosis(signal)
    skewness = skew(signal)

    # Append features to the list
    features.append([mean, std, max_value, min_value, entropy_value, kurtosis_value,
skewness])

  return features

# Extract features from the current signals
normal_features = extract_features(normal_data)
faulty_features = extract_features(faulty_data)

# Create labels for normal and faulty data
normal_labels = np.zeros(len(normal_features))
faulty_labels = np.ones(len(faulty_features))

# Combine features and labels
all_features = np.concatenate((normal_features, faulty_features), axis=0)
all_labels = np.concatenate((normal_labels, faulty_labels), axis=0)

# Split the data into training and testing sets
X_train, X_test, y_train, y_test = train_test_split(all_features, all_labels, test_size=0.2,
random_state=42)

# Train an SVM classifier
model = SVC(kernel='rbf', C=1.0, random_state=42)
model.fit(X_train, y_train)

# Predict labels for the test set
y_pred = model.predict(X_test)
```

```
# Calculate accuracy
accuracy = accuracy_score(y_test, y_pred)
print("Accuracy:", accuracy)
```

19 Example: Bearing RUL prediction

```
import numpy as np
import torch
import torch.nn as nn
import torch.optim as optim
from torch.utils.data import Dataset, DataLoader
from sklearn.preprocessing import MinMaxScaler
from sklearn.model_selection import train_test_split

# Define the RUL prediction model
class RULPredictionModel(nn.Module):
    def __init__(self, input_dim, hidden_dim, output_dim):
        super(RULPredictionModel, self).__init__()
        self.hidden_dim = hidden_dim
        self.rnn = nn.GRU(input_dim, hidden_dim, batch_first=True)
        self.fc = nn.Linear(hidden_dim, output_dim)

    def forward(self, x):
        _, hidden = self.rnn(x)
        output = self.fc(hidden[-1])
        return output.squeeze(1)

# Define a custom dataset for RUL prediction
class RULDataset(Dataset):
    def __init__(self, data, targets):
```

```
        self.data = data
        self.targets = targets

    def __len__(self):
        return len(self.data)

    def __getitem__(self, idx):
        return self.data[idx], self.targets[idx]

# Sample data
# Assuming you have preprocessed data: X_train, X_test, y_train, y_test

# Normalize the data
scaler = MinMaxScaler()
X_train = scaler.fit_transform(X_train)
X_test = scaler.transform(X_test)

# Convert the data to PyTorch tensors
X_train = torch.tensor(X_train).float()
X_test = torch.tensor(X_test).float()
y_train = torch.tensor(y_train).float()
y_test = torch.tensor(y_test).float()

# Define the model hyperparameters
input_dim = X_train.shape[1]  # Input dimension
hidden_dim = 64  # Hidden dimension of the GRU
output_dim = 1  # Output dimension

# Create the RUL prediction model
model = RULPredictionModel(input_dim, hidden_dim, output_dim)
```

```
# Define the loss function and optimizer
criterion = nn.MSELoss()
optimizer = optim.Adam(model.parameters(), lr=0.001)

# Create dataloaders for training and testing
train_dataset = RULDataset(X_train, y_train)
train_dataloader = DataLoader(train_dataset, batch_size=64, shuffle=True)
test_dataset = RULDataset(X_test, y_test)
test_dataloader = DataLoader(test_dataset, batch_size=64, shuffle=False)

# Training loop
num_epochs = 100
for epoch in range(num_epochs):
    model.train()
    for batch_inputs, batch_targets in train_dataloader:
        optimizer.zero_grad()
        outputs = model(batch_inputs.unsqueeze(1))
        loss = criterion(outputs, batch_targets.unsqueeze(1))
        loss.backward()
        optimizer.step()

    model.eval()
    with torch.no_grad():
        test_loss = 0
        for batch_inputs, batch_targets in test_dataloader:
            outputs = model(batch_inputs.unsqueeze(1))
            batch_loss = criterion(outputs, batch_targets.unsqueeze(1))
            test_loss += batch_loss.item() * batch_inputs.size(0)

        test_loss /= len(test_dataset)
        print(f"Epoch {epoch+1}/{num_epochs}, Test Loss: {test_loss:.4f}")
```

20 Horizontal federated learning

```
import torch
import torch.nn as nn
import torch.optim as optim
import torch.utils.data as data
from torchvision import datasets, transforms

# Define the model architecture
class Net(nn.Module):
    def __init__(self):
        super(Net, self).__init__()
        self.fc1 = nn.Linear(784, 128)
        self.relu = nn.ReLU()
        self.fc2 = nn.Linear(128, 10)
        self.softmax = nn.LogSoftmax(dim=1)

    def forward(self, x):
        x = x.view(-1, 784)
        x = self.fc1(x)
        x = self.relu(x)
        x = self.fc2(x)
        x = self.softmax(x)
        return x

# Define the training function
def train(model, train_loader, optimizer, criterion, epoch):
    model.train()
    for batch_idx, (data, target) in enumerate(train_loader):
        optimizer.zero_grad()
```

```
        output = model(data)
        loss = criterion(output, target)
        loss.backward()
        optimizer.step()
        if batch_idx % 100 == 0:
            print('Train Epoch: {} [{}/{} ({:.0f}%)]\tLoss: {:.6f}'.format(
                epoch, batch_idx * len(data), len(train_loader.dataset),
                100. * batch_idx / len(train_loader), loss.item()))

# Define the test function
def test(model, test_loader):
    model.eval()
    test_loss = 0
    correct = 0
    with torch.no_grad():
        for data, target in test_loader:
            output = model(data)
            test_loss += criterion(output, target).item()
            pred = output.argmax(dim=1, keepdim=True)
            correct += pred.eq(target.view_as(pred)).sum().item()

    test_loss /= len(test_loader.dataset)
    accuracy = 100. * correct / len(test_loader.dataset)
    print('\nTest set: Average loss: {:.4f}, Accuracy: {}/{} ({:.2f}%)\n'.format(
        test_loss, correct, len(test_loader.dataset), accuracy))

# Set the random seed for reproducibility
torch.manual_seed(1234)

# Define the device
device = torch.device('cuda' if torch.cuda.is_available() else 'cpu')
```

```
# Define the hyperparameters
batch_size = 64
learning_rate = 0.01
num_epochs = 10

# Load the MNIST dataset
train_dataset = datasets.MNIST(
    './data', train=True, download=True,
    transform=transforms.Compose([
        transforms.ToTensor(),
        transforms.Normalize((0.1307,), (0.3081,))
    ])
)

# Create data loaders for each participant
train_loader_1 = data.DataLoader(
    train_dataset, batch_size=batch_size, shuffle=True, num_workers=2
)

train_loader_2 = data.DataLoader(
    train_dataset, batch_size=batch_size, shuffle=True, num_workers=2
)

# Create the model and optimizer
model = Net().to(device)
optimizer = optim.SGD(model.parameters(), lr=learning_rate)

# Define the loss function
criterion = nn.CrossEntropyLoss()
```

```
# Perform federated learning
for epoch in range(num_epochs):
    train(model, train_loader_1, optimizer, criterion, epoch)
    train(model, train_loader_2, optimizer, criterion, epoch)

# Test the federated model
test_dataset = datasets.MNIST(
    './data', train=False, download=True,
    transform=transforms.Compose([
        transforms.ToTensor(),
        transforms.Normalize((0.1307,), (0.3081,))
    ])
)
test_loader = data.DataLoader(test_dataset, batch_size=batch_size, shuffle=True)
test(model, test_loader)
```

21 Vertical Federated Learning

```
import torch
import torch.nn as nn
import torch.optim as optim
import torch.utils.data as data

# Define the model architecture
class Net(nn.Module):
    def __init__(self):
        super(Net, self).__init__()
        self.fc1 = nn.Linear(2, 1)
        self.sigmoid = nn.Sigmoid()

    def forward(self, x):
```

```
        x = self.fc1(x)
        x = self.sigmoid(x)
        return x

# Define the training function
def train(model, train_loader, optimizer, criterion, epoch):
    model.train()
    for batch_idx, (data, target) in enumerate(train_loader):
        optimizer.zero_grad()
        output = model(data)
        loss = criterion(output, target)
        loss.backward()
        optimizer.step()
        if batch_idx % 10 == 0:
            print('Train Epoch: {} [{}/{} ({:.0f}%)]\tLoss: {:.6f}'.format(
                epoch, batch_idx * len(data), len(train_loader.dataset),
                100. * batch_idx / len(train_loader), loss.item()))

# Set the random seed for reproducibility
torch.manual_seed(1234)

# Define the hyperparameters
batch_size = 64
learning_rate = 0.01
num_epochs = 10

# Generate synthetic data for the two participants
# Participant 1 owns feature 1 and target
X_train_1 = torch.rand(100, 1)
y_train_1 = torch.randint(0, 2, (100,))
```

```
# Participant 2 owns feature 2 and target
X_train_2 = torch.rand(100, 1)
y_train_2 = torch.randint(0, 2, (100,))

# Create data loaders for each participant
train_dataset_1 = data.TensorDataset(X_train_1, y_train_1)
train_loader_1 = data.DataLoader(train_dataset_1,batch_size=batch_size,shuffle=True)

train_dataset_2 = data.TensorDataset(X_train_2,y_train_2)
train_loader_2 = data.DataLoader(train_dataset_2,batch_size=batch_size,shuffle=True)

# Create the global model and optimizer
model = Net()
optimizer = optim.SGD(model.parameters(), lr=learning_rate)

# Define the loss function
criterion = nn.BCELoss()

# Perform vertical federated learning
for epoch in range(num_epochs):
   train(model, train_loader_1, optimizer, criterion, epoch)
   train(model, train_loader_2, optimizer, criterion, epoch)

# Test the federated model
X_test = torch.rand(10, 2)  # Test data with both features
test_dataset = data.TensorDataset(X_test)
test_loader = data.DataLoader(test_dataset, batch_size=batch_size)

model.eval()
with torch.no_grad():
   for data in test_loader:
```

```
    output = model(data[0])
    print(output)
```

22 Federated transfer learning

```
import torch
import torch.nn as nn
import torch.optim as optim
import torch.utils.data as data
from torchvision import datasets, transforms, models

# Define the model architecture
class TransferNet(nn.Module):
    def __init__(self, num_classes):
        super(TransferNet, self).__init__()
        self.model = models.resnet18(pretrained=True)
        num_features = self.model.fc.in_features
        self.model.fc = nn.Linear(num_features, num_classes)

    def forward(self, x):
        x = self.model(x)
        return x

# Define the training function
def train(model, train_loader, optimizer, criterion, epoch):
    model.train()
    for batch_idx, (data, target) in enumerate(train_loader):
        optimizer.zero_grad()
        output = model(data)
        loss = criterion(output, target)
        loss.backward()
```

```
        optimizer.step()
        if batch_idx % 10 == 0:
            print('Train Epoch: {} [{}/{} ({:.0f}%)]\tLoss: {:.6f}'.format(
                epoch, batch_idx * len(data), len(train_loader.dataset),
                100. * batch_idx / len(train_loader), loss.item()))

# Set the random seed for reproducibility
torch.manual_seed(1234)

# Define the hyperparameters
batch_size = 64
learning_rate = 0.001
num_epochs = 10
num_classes = 10

# Load the pre-trained model
pretrained_model = models.resnet18(pretrained=True)

# Create data loaders for each participant
train_dataset_1 = datasets.CIFAR10(
    './data', train=True, download=True,
    transform=transforms.Compose([
        transforms.RandomCrop(32, padding=4),
        transforms.RandomHorizontalFlip(),
        transforms.ToTensor(),
        transforms.Normalize((0.5, 0.5, 0.5), (0.5, 0.5, 0.5))
    ])
)

train_loader_1=data.DataLoader(train_dataset_1,batch_size=batch_size,shuffle=True)
```

```
train_dataset_2 = datasets.CIFAR10(
    './data', train=True, download=True,
    transform=transforms.Compose([
        transforms.RandomCrop(32, padding=4),
        transforms.RandomHorizontalFlip(),
        transforms.ToTensor(),
        transforms.Normalize((0.5, 0.5, 0.5), (0.5, 0.5, 0.5))
    ])
)

train_loader_2 = data.DataLoader(train_dataset_2, batch_size=batch_size,shuffle=True)

# Create the global model and optimizer
model = TransferNet(num_classes)
model.load_state_dict(pretrained_model.state_dict())  # Initialize the global model
with the pre-trained weights

optimizer = optim.Adam(model.parameters(), lr=learning_rate)

# Define the loss function
criterion = nn.CrossEntropyLoss()

# Perform federated transfer learning
for epoch in range(num_epochs):
    train(model, train_loader_1, optimizer, criterion, epoch)
    train(model, train_loader_2, optimizer, criterion, epoch)

# Test the federated model
test_dataset = datasets.CIFAR10(
    './data', train=False, download=True,
    transform=transforms.Compose([
```

```
        transforms.ToTensor(),
        transforms.Normalize((0.5, 0.5, 0.5), (0.5, 0.5, 0.5))
    ])
)

test_loader = data.DataLoader(test_dataset, batch_size=batch_size, shuffle=True)

model.eval()
correct = 0
total = 0
with torch.no_grad():
    for images, labels in test_loader:
        images = images.to(device)
        labels = labels.to(device)
        outputs = model(images)
        _, predicted = torch.max(outputs.data, 1)
        total += labels.size(0)
        correct += (predicted == labels).sum().item()

accuracy = 100 * correct / total
print('Accuracy on the test set: {:.2f}%'.format(accuracy))
```

References

[1] H. Henao, G. Capolino, M. F. Cabanas, *et al.*, "Trends in fault diagnosis for electrical machines: a review of diagnostic techniques," *IEEE Industrial Electronics Magazine*, vol. 8, no. 2, pp. 31–42, 2014.

[2] S. S. S. R. Sarathbabu Duvvuri, "Rotor inter-turn short-circuit fault diagnostics relevant modeling of slip-ring induction motor," in *2018 IEEE 8th Power India International Conference (PIICON)*, 2018, pp. 1–6.

[3] J. Antonino-Daviu, S. Aviyente, E. G. Strangas and M. Riera-Guasp, "Scale invariant feature extraction algorithm for the automatic diagnosis of rotor asymmetries in induction motors," *IEEE Transactions on Industrial Informatics*, vol. 9, no. 1, pp. 100–108, 2013.

[4] Y. Kim, Y. Youn, D. Hwang, J. Sun and D. Kang, "High-resolution parameter estimation method to identify broken rotor bar faults in induction motors," *IEEE Transactions on Industrial Electronics*, vol. 60, no. 9, pp. 4103–4117, 2013.

[5] L. M. Contreras-Medina, R. J. Romero-Troncoso, E. Cabal-Yepez, J. J. Rangel-Magdaleno and J. R. Millan-Almaraz, "FPGA-based multiple-channel vibration analyzer for industrial applications in induction motor failure detection," *IEEE Transactions on Instrumentation and Measurement*, vol. 59, no. 1, pp. 63–72, 2010.

[6] S. X. Ding, P. Zhang, S. Yin and E. L. Ding, "An integrated design framework of fault-tolerant wireless networked control systems for industrial automatic control applications," *IEEE Transactions on Industrial Informatics*, vol. 9, no. 1, pp. 462–471, 2013.

[7] A. Khandekar and M. A. Singhal, "Bearing fault detection in induction motor using fast Fourier transform," in *Proc. IEEE Int. Conf. Adv. Res. Eng. Technol.*, Feb. 2013, pp. 190–194.

[8] A. Soualhi, G. Clerc and H. Razik, "Detection and diagnosis of faults in induction motor using an improved artificial ant clustering technique," *IEEE Transactions on Industrial Electronics*, vol. 60, no. 9, pp. 4053–4062, 2013.

[9] F. Immovilli, C. Bianchini, M. Cocconcelli, A. Bellini and R. Rubini, "Bearing fault model for induction motor with externally induced vibration," *IEEE Transactions on Industrial Electronics*, vol. 60, no. 8, pp. 3408–3418, 2013.

[10] Z. Wang and C. S Chang, "Online fault detection of induction motors using frequency domain independent components analysis," in *Proc. IEEE Int. Symp. Ind. Electron*, Jun. 2011, pp. 2132–2137.

[11] Z. Wang, C. S. Chang and Y. Zhang, "A feature based frequency domain analysis algorithm for fault detection of induction motors," in *Proc. 6th IEEE Conf. Ind. Electron. Appl.*, Jun. 2011, pp. 27–32.

[12] M. Seera, C. P. Lim, S. Nahavandi and C. K. Loo, "Condition monitoring of induction motors: a review and an application of an ensemble of hybrid intelligent models," *Expert Systems with Applications*, vol. 41, no. 10, pp. 4891–4903, 2014.

[13] M. Drif and A. J. M Cardoso, "Stator fault diagnostics in squirrel cage three-phase induction motor drives using the instantaneous active and reactive power signature analyses," *IEEE Transactions on Industrial Informatics*, pp. 1348–1360, 2014.

[14] E. H. El Bouchikhi, V. Choqueuse and M. Benbouzid, "Induction machine faults detection using stator current parametric spectral estimation," *Mechanical Systems and Signal Processing*, vols. 52–53, pp. 447–464, 2015.

[15] K. A. Loparo, M. L. Adams and M. F. Abdel-Magied, *et al.*, "Fault detection and diagnosis of rotating machinery," *IEEE Transactions on Industrial Electronics*, vol. 47, no. 5, pp. 1005–1014, 2000.

[16] T. W. S. Chow and H. Z. Tan, "HOS-based nonparametric and parametric methodologies for machine fault detection," *IEEE Transactions on Industrial Electronics*, vol. 47, no. 5, pp. 1051–1059, 2000.

[17] M. E. H. Benbouzid, "A review of induction motors signature analysis as a medium for faults detection," *IEEE Transactions on Industrial Electronics*, vol. 47, no. 5, pp. 984–993, 2000.

[18] P. Priyanga S, K. Krithivasan, Pravinraj S and S. Sriram V S, "Detection of cyberattacks in industrial control systems using enhanced principal component analysis and hypergraph-based convolution neural network (EPCA-HG-CNN)," *IEEE Transactions on Industry Applications*, vol. 56, no. 4, pp. 4394–4404, 2020.

[19] M. Kaminski, C. T. Kowalski and T. Orlowska-Kowalska, "Application of radial basis neural networks for the rotor fault detection of the induction motor," in *2011 IEEE EUROCON – International Conference on Computer as a Tool*, 2011, pp. 1–4.

[20] E. M. Tag Eldin, H. R. Emara, E.M. Aboul-Zahab and S. S. Refaat, "Monitoring and diagnosis of external faults in three phase induction motors using artificial neural network," in *2007 IEEE Power Engineering Society General Meeting*, 2007 1-4244-1298-2007.

[21] S. Premrudee preechacharn, T. Utthiyuung, K. Kruepengkul, P. Puongkaew, "Induction motor fault detection and diagnosis using supervised and unsupervised neural networks," in *2002 IEEE International Conference on Industrial Technology, 2002. IEEE ICIT '02*, Bangkok, Thailand, 2002 0-7803-7657.91021-2002.

[22] R. Natarajan, "Failure identification of induction motors by sensing unbalanced stator currents," *IEEE Transactions on Energy Conversion*, vol. 4, no. 4, 1989.

[23] S. Aksoy and A. Mühürcü, "Recurrent neural network based nonlinear state estimation for induction motors," *Journals of International Review of Electrical Engineering*, vol. 6, no. 2, 2011 Papers Part A.

[24] M. Fernandez-Temprano, P. E. Gardel-Sotomayor, O. Duque-Perez and D. Morinigo-Sotelo, "Broken bar condition monitoring of an induction motor under different supplies using a linear discriminant analysis," in *Proc. 9th IEEE Int. Symp. Diagnostics Electr. Mach., Power Electron. Drives*, Aug. 2013, pp. 162–168.

[25] J. Seshadrinath, B. Singh and B. K. Panigrahi, "Investigation of vibration signatures for multiple fault diagnosis in variable frequency drives using complex wavelets," *IEEE Transactions on Power Electronics*, vol. 29, no. 2, pp. 936–945, 2014.

[26] W. S. Gongora, H. V. D. Silva, A. Goedtel, W. F. Godoy and S. A. Q. da Silva, "Neural approach for bearing fault detection in three phase induction motors," in *Proc. 9th IEEE Int. Symp. Diagnostics Electr. Mach., Power Electron. Drives*, Aug. 2013, pp. 566–572.

[27] O. Duque-Perez, L.-A. Garcia-Escudero, D. Morinigo-Sotelo, P.-E. Gardel and M. Perez-Alonso, "Analysis of fault signatures for the diagnosis of induction motors fed by voltage source inverters using ANOVA and additive models," *Electric Power Systems Research*, vol. 121, pp. 1–13, 2015.

[28] M. Unal, M. Onat, M. Demetgul and H. Kucuk, "Fault diagnosis of rolling bearings using a genetic algorithm optimized neural network," *Measurement: Journal of the International Measurement Confederation*, vol. 58, pp. 187–196, 2014.

[29] D. Yu, J. Cheng and Y. Yang, "Application of EMD method and Hilbert spectrum to the fault diagnosis of roller bearings," *Mechanical Systems and Signal Processing*, vol. 19, no. 3, pp. 259–270, 2005.

[30] Z. K. Peng, P. W. Tse and F. L. Chu, "An improved Hilbert-Huang transform and its application in vibration signal analysis," *Journal of Sound and Vibration*, vol. 286, pp. 187–205, 2005.

[31] G. K. Singh and S. A. K. S Ahmed, "Vibration signal analysis using wavelet transform for isolation and identification of electrical faults in induction machine," *Electric Power Systems Research*, vol. 68, pp. 119–136, 2004.

[32] V. Iglesias, J. Grajal, M. A. Sánchez and M. López-Vallejo, "Implementation of a real-time spectrum analyzer on FPGA platforms," *IEEE Transactions on Instrumentation and Measurement*, vol. 64, no. 2, pp. 338–355, 2015.

[33] G. Betta, C. Liguori, A. Paolillo and A. Pietrosanto, "A DSP-based FFT-analyzer for the fault diagnosis of rotating machine based on vibration analysis," *IEEE Transactions on Instrumentation and Measurement*, vol. 51, no. 6, 2002.

[34] V.K. Rai and A.R. Mohanty, "Bearing fault diagnosis using FFT of intrinsic mode functions in Hilbert-Huang transform," *Mechanical Systems and Signal Processing*, vol. 21, no. 8, pp. 2607–2615, 2007.

[35] Y. Qin, S. Qin and Y. Mao, "Research on iterated Hilbert transform and its application in mechanical fault diagnosis," *Mechanical Systems and Signal Processing*, vol. 22, pp. 1967–1980, 2008.

[36] N. Mohanti, *Signal Processing Signal Filtering Detection*. New York: Van Nostrand, 1987.

[37] A. F. Atiya and A. G. Palos, "New results on recurrent network training: unifying the algorithms and accelerating convergence," *IEEE Transactions on Neural Networks*, vol. 13, pp. 765–786, 2000.

[38] H. Su and K. T. Chong, "Induction machine condition monitoring using neural network modeling," *IEEE Transactions on Industrial Electronics*, vol. 54, no. 1, pp. 241–249, 2007.

[39] R. Jiang, S. Liu, Y. Tang and Y. Liu, "A novel method of fault diagnosis for rolling element bearings based on the accumulated envelope spectrum of the wavelet packet," *Journal of Vibration and Control*, vol. 21, no. 8, pp. 1580–1593, 2015.

[40] J. Chen, Z. Li, J. Pan, and C. Gaige, "Wavelet transform based on inner product in fault diagnosis of rotating machinery: a review," *Mechanical Systems and Signal Processing*, vols. 70–71, pp. 1–35, 2016.

[41] R. G. C. Cunha, E. T. da Silva Jr and C. M. de Sá Medeiros, "Machine learning and multiresolution decomposition for embedded applications to detect short-circuit in induction motors," *Computers in Industry*, vol. 129, pp. 103461, 2021.

[42] B. M. Ebrahimi, J. Faiz, S. Lotfi-Fard and P. Pillay, "Novel indices for broken rotor bars fault diagnosis in induction motors using wavelet transform," *Mechanical Systems and Signal Processing*, vol. 30, no. 7, pp. 131–145, 2012.

[43] R. Yan, R. X. Gao and X. Chen, "Wavelets for fault diagnosis of rotary machines: a review with applications," *Signal Process*, vol. 96, pp. 1–15, 2014.

[44] F. Ding, X. Zhang, W. Wu and Y. Wang, "A fault feature extraction method of motor bearing using improved LCD," *IEEE Access*, pp. 220973–220979, 2020.

[45] N. Rehman, M. M. Khan, M. I. Sohaib, M. Jehanzaib, S. Ehsan and K. McDonald-Maier, "Image fusion using multivariate and multidimensional EMD," in *2014 IEEE International Conference on Image Processing (ICIP)*, 2014, pp. 5112–5116.

[46] J. H. Ahn, D.-H. Kwak and B.-H. Koh, "Fault detection of a roller-bearing system through the EMD of a wavelet denoised signal," *Sensors*, vol. 14, no. 8, pp. 15022–15038, 2014.

[47] J. Gilles, "Empirical wavelet transform," *IEEE Transactions on Signal Processing*, vol. 61, no. 16, pp. 3999–4010, 2013.

[48] K. Dragomiretskiy and D. Zosso, "Variational mode decomposition," *IEEE Transactions on Signal Processing*, vol. 62, no. 3, pp. 531–544, 2014.

[49] X. B. Wang, Z. X. Yang and X. A. Yan, "Novel particle swarm optimization-based variational mode decomposition method for the fault

diagnosis of complex rotating machinery," *IEEE/ASME Transactions on Mechatronics*, vol. 23, no. 1, pp. 68–79, 2018.

[50] J. S. Cheng, J. D. Zheng and Y. Yang, "A nonstationary signal analysis approach—the local characteristic-scale decomposition method," *Journal of Vibration Engineering*, vol. 25, no. 2, pp. 215–220, 2012.

[51] Z. Wu, J. Cheng and Y. Yang, "Autonomous dense local feature scale decomposition method and its application," *Journal of Mechanical Engineering*, vol. 6, no. 11, pp. 1464–1470, 2015.

[52] L. Chenjie, W. Bin and T. Tao, "Blind estimation algorithm of frequency hopping signal parameters using local feature scale decomposition," *Signal Process*, vol. 5, no. 3, pp. 308–313, 2015.

[53] Y. Huang, C. H. Chen and C. J. Huang, "Motor fault detection and feature extraction using RNN-based variational autoencoder," *IEEE Access*, vol. 7, pp. 139086–139096, 2019.

[54] V. Choqueuse, M. E. H. Benbouzid, Y. Amirat and S. Turri, "Diagnosis of three-phase electrical machines using multidimensional demodulation techniques," *IEEE Transactions on Industrial Electronics*, vol. 59, no. 4, pp. 2014–2023, 2012.

[55] L. A. O Martins, F. L. C. Pádua and P. E. M. Almeida, "Automatic detection of surface defects on rolled steel using computer vision and artificial neural networks," in *Proc. IEEE IECON*, Nov. 2010, pp. 1081–1086.

[56] M. I. Chacon-Murguia and S. Gonzalez-Duarte, "An adaptive neural-fuzzy approach for object detection in dynamic backgrounds for surveillance systems," *IEEE Transactions on Industrial Electronics*, vol. 59, no. 8, pp. 3286–3298, 2012.

[57] M. D. Prieto, G. Cirrincione, A. G. Espinosa, J. A. Ortega and H. Henao, "Bearing fault detection by a novel condition-monitoring scheme based on statistical-time features and neural networks," *IEEE Transactions on Industrial Electronics*, vol. 60, no. 8, 2013.

[58] X. F. St-Onge, J. Cameron, S. Saleh and E. J. Scheme, "A symmetrical component feature extraction method for fault detection in induction machines," *IEEE Transactions on Industrial Electronics*, vol. 66, no. 9, pp. 7281–7289, 2019.

[59] C. Chen, B. Zhang, G. Vachtsevanos and M. Orchard, "Machine condition prediction based on adaptive neuro-fuzzy and high-order particle filtering," *IEEE Transactions on Industrial Electronics*, vol. 58, no. 9, pp. 4353–4364, 2011.

[60] X. An and H. Zeng, "Fault diagnosis method for spherical roller bearing of wind turbine based on variational mode decomposition and singular value decomposition," *Journal of Vibroengineering*, vol. 18, no. 6, pp. 3548–3556, 2016.

[61] S. W. Fei, "Fault diagnosis of bearing based on wavelet packet transform phase space reconstruction-singular value decomposition and SVM classifier," *Arabian Journal for Science and Engineering*, vol. 42, no. 5, pp. 1967–1975, 2017.

[62] D. T. Hoang and H. J. Kang, "A motor current signal-based bearing fault diagnosis using deep learning and information fusion," *IEEE Transactions on Instrumentation and Measurement*, vol. 69, no. 6, pp. 3325–3333, 2020.

[63] X. Zhang, B. Wang and X. Chen, "Intelligent fault diagnosis of roller bearings with multivariable ensemble-based incremental support vector machine," *Knowledge-Based Systems*, vol. 89, pp. 56–85, 2015.

[64] Q. Hu, X. Si, A. Qin, Y. Lv and Q. Zhang, "Machinery fault diagnosis scheme using redefined dimensionless indicators and mRMR feature selection," *IEEE Access*, vol. 8, pp. 40313–40326, 2020.

[65] J. Zheng, H. Pan and J. Cheng, "Rolling bearing fault detection and diagnosis based on composite multiscale fuzzy entropy and ensemble support vector machines," *Mechanical Systems and Signal Processing*, vol. 85, pp. 746–759, 2017.

[66] S. Das, P. Purkait, C. Koley and S. Chakravorti, "Performance of a load-immune classifier for robust identification of minor faults in induction motor stator winding," *IEEE Transactions on Dielectrics and Electrical Insulation*, vol. 21, no. 1, pp. 33–44, 2014.

[67] W. Fengqi and G. Meng, "Compound rub malfunctions feature extraction based on full-spectrum cascade analysis and SVM," *Mechanical Systems and Signal Processing*, vol. 20, no. 8, pp. 2007–2021, 2006.

[68] V. N. Ghate and S. V. Dudul, "Cascade neural-network-based fault classifier for three-phase induction motor," *IEEE Transactions on Industrial Electronics*, vol. 58, no. 5, pp. 1555–1563, 2011.

[69] S. Zhang, X. Li, M. Zong, X. Zhu and R. Wang, "Efficient kNN classification with different numbers of nearest neighbors," *IEEE Transactions on Neural Networks and Learning Systems*, vol. 29, no. 5, pp. 1774–1785, 2018.

[70] A. Recioui, B. Benseghier and H. Khalfallah, "Power system fault detection, classification and location using the K-nearest neighbors," in *2015 4th International Conference on Electrical Engineering, ICEE 2015*, 2016.

[71] Q. Yao, J. Wang, L. Yang, H. Su and G. Zhang, "A fault diagnosis method of engine rotor based on Random Forests," in *2016 IEEE International Conference on Prognostics and Health Management, ICPHM 2016*, 2016, pp. 0–3.

[72] R. K. Patel and V. K. Giri, "Feature selection and classification of mechanical fault of an induction motor using random forest classifier," *Perspectives in Science*, vol. 8, pp. 334–337, 2016.

[73] M. Seera, C. P. Lim and C. K. Loo, "Motor fault detection and diagnosis using a hybrid FMM-CART model with online learning," *Journal of Intelligent Manufacturing*, vol. 27, no. 6, pp. 1273–1285, 2016.

[74] T. Ince, S. Kiranyaz, L. Eren, M. Askar and M. Gabbouj, "Real-time motor fault detection by 1-D convolutional neural networks," *IEEE Transactions on Industrial Electronics*, vol. 63, no. 11, pp. 7067–7075, 2016.

[75] Y. Lei, F. Jia, J. Lin, S. Xing and S. X. Ding, "An intelligent fault diagnosis method using unsupervised feature learning towards mechanical big data," *IEEE Transactions on Industrial Electronics*, vol. 63, no. 5, pp. 3137–3147, 2016.

[76] H. Chen, B. Jiang, T. Zhang and N. Lu, "Data-driven and deep learning-based detection and diagnosis of incipient faults with application to electrical traction systems," *Neurocomputing*, vol. 396, pp. 429–437, 2020.
[77] S. Shao, R. Yan, Y. Lu, P. Wang and R. X. Gao, "DCNN-Based multi-signal induction motor fault diagnosis," *IEEE Transactions on Instrumentation and Measurement*, vol. 69, no. 6, pp. 2658–2669, 2020.
[78] K. Liang, N. Qin, D. Huang, Y. Fu and L. Hu, "Convolutional recurrent neural network for fault diagnosis of high-speed train bogie," *Complexity*, vol. 2018, p. 4501952, 2018.
[79] Z. Chen, K. Gryllias and W. Li, "Mechanical fault diagnosis using convolutional neural networks and extreme learning machine," *Mechanical Systems and Signal Processing*, vol. 133, p. 106272, 2019.
[80] A. Shenfield and M. Howarth, "A novel deep learning model for the detection and identification of rolling element-bearing faults," *Sensors (Switzerland)*, vol. 20, no. 18, pp. 1–24, 2020.
[81] W. Zhang, G. Peng, C. Li, Y. Chen and Z. Zhang, "A new deep learning model for fault diagnosis with good anti-noise and domain adaptation ability on raw vibration signals," *Sensors (Switzerland)*, vol. 17, no. 2, 2017.
[82] W. Zhang, C. Li, G. Peng, Y. Chen and Z. Zhang, "A deep convolutional neural network with new training methods for bearing fault diagnosis under noisy environment and different working load," *Mechanical Systems and Signal Processing*, vol. 100, pp. 439–453, 2018.
[83] J. S. L. Senanayaka, H. Van Khang and K. G. Robbersmyr, "Towards self-supervised feature learning for online diagnosis of multiple faults in electric powertrains," *IEEE Transactions on Industrial Informatics*, vol. 17, no. 6, pp. 3772–3781, 2021.
[84] L. Wen, X. Li, L. Gao and Y. Zhang, "A new convolutional neural network-based data-driven fault diagnosis method," *IEEE Transactions on Industrial Electronics*, vol. 65, no. 7, pp. 5990–5998, 2018.
[85] M. Jalayer, C. Orsenigo and G. Vercellis, "Fault detection and diagnosis for rotating machinery: a model based on convolutional LSTM, fast Fourier and continuous wavelet transforms," *Computers in Industry*, vol. 125, p. 103378, 2021.
[86] R. Zhao, R. Yan, J. Wang and K. Mao, K, "Learning to monitor machine health with convolutional Bi-directional LSTM networks," *Sensors (Switzerland)*, vol. 17, no. 2, 2017.
[87] T. Wong and Z. Luo, "Recurrent auto-encoder model for large-scale industrial sensor signal analysis," in *Proc. Int. Conf. Eng. Appl. Neural Netw.*, Jul. 2018, pp. 203–216.
[88] L. D. Xu and L. Duan, "Big data for cyber physical systems in industry 4.0: a survey," *Enterprise Information Systems*, vol. 13, no. 2, pp. 148–169, 2019.
[89] Y. Ma, K. Liu, Z. Guan, X. Xu, X. Qian and H. Bao, "Background augmentation generative adversarial networks (BAGANs): Effective data generation based on GAN-augmented 3D synthesizing," *Symmetry*, vol. 10, no. 12, p. 734, 2018.

[90] L. Deng, "A tutorial survey of architectures, algorithms, and applications for deep learning," *APSIPA Trans. Signal Inf. Process*, vol. 3, 2014.

[91] K. Cheng, R. Tahir, L. K. Eric and M. Li, "An analysis of generative adversarial networks and variants for image synthesis on MNIST dataset," *Multimedia Tools and Applications*, vol. 79, nos. 19–20, pp. 13725–13752, 2020.

[92] H. Apaydin, H. Feizi, M. T. Sattari, M. S. Colak, S. Shamshirband and K.-W. Chau, "Comparative analysis of recurrent neural network architectures for reservoir inflow forecasting," *Water*, vol. 12, no. 5, p. 1500, 2020.

[93] C. Zhang, C. Xiong and L. Wang, "A research on generative adversarial networks applied to text generation," in *Proc. 14th Int. Conf. Comput. Sci. Educ. (ICCSE)*, Aug. 2019, pp. 913–917.

[94] Y. O. Lee, J. Jo and J. Hwang, "Application of deep neural network and generative adversarial network to industrial maintenance: a case study of induction motor fault detection," in *IEEE Int. Conf.*, 2017, pp. 3248–3253.

[95] Z. Wang, J. Wang and Y. Wang, "An intelligent diagnosis scheme based on generative adversarial learning deep neural networks and its application to planetary gearbox fault pattern recognition," *Neurocomputing*, vol. 310, pp. 213–222, 2018.

[96] L. Wen, X. Li, L. Gao and Y. Zhang, "A generative adversarial network-based intelligent fault diagnosis method for rotating machinery under small sample size conditions," *IEEE Access*, vol. 7, pp. 149736–149749, 2019.

[97] M. Gan, C. Wang and C. Zhu, "Construction of hierarchical diagnosis network based on deep learning and its application in the fault pattern recognition of rolling element bearings," *Mechanical Systems and Signal Processing*, vol. 72–73, pp. 92–104, 2016.

[98] H. Shao, H. Jiang, H. Zhang, W. Duan, T. Liang and S. Wu, "Rolling bearing fault feature learning using improved convolutional deep belief network with compressed sensing," *Mechanical Systems and Signal Processing*, vol. 100, pp. 743–765, 2018.

[99] M. Ma, C. Sun and X. Chen, "Discriminative deep belief networks with ant colony optimization for health status assessment of machine," *IEEE Transactions on Instrumentation and Measurement*, vol. 66, no. 12, pp. 3115–3125, 2017.

[100] T. Zhang, W. Wang, H. Ye, D. Huang, H. Zhang and M. Li, "Fault detection for ironmaking process based on stacked denoising autoencoders," in *Proceedings of the American Control Conference*, Jul. 2016, pp. 3261–3267.

[101] C. Lu, Z. Y. Wang, W. L. Qin and J. Ma, "Fault diagnosis of rotary machinery components using a stacked denoising autoencoder-based health state identification," *Signal Processing*, vol. 130, pp. 377–388, 2017.

[102] H. Lee, Y. Kim and C.O. Kim, "A deep learning model for robust wafer fault monitoring with sensor measurement noise," *IEEE Transactions on Semiconductor Manufacturing*, vol. 30, pp. 23–31, 2017.

[103] F. Jia, Y. Lei, L. Guo, J. Lin and S. Xing, "A neural network constructed by deep learning technique and its application to intelligent fault diagnosis of machines," *Neurocomputing*, vol. 272, pp. 619–628, 2018.

[104] W. Mao, W. Feng, Y. Liu, D. Zhang and X. Liang, "A new deep auto-encoder method with fusing discriminant information for bearing fault diagnosis," *Mechanical Systems and Signal Processing*, vol. 150, p. 107233, 2021.

[105] F. Jia, Y. Lei, J. Lin, X. Zhou and N. Lu, "Deep neural networks: a promising tool for fault characteristic mining and intelligent diagnosis of rotating machinery with massive data," *Mechanical Systems and Signal Processing*, vols. 72–73, pp. 303–315, 2016.

[106] Z. Chen and W. Li, "Multisensor feature fusion for bearing fault diagnosis using sparse autoencoder and deep belief network," *IEEE Transactions on Instrumentation and Measurement*, vol. 66, no. 7, pp. 1693–1702, 2017.

[107] J. An, P. Ai, C. Liu, S. Xu and D. Liu, "Deep clustering bearing fault diagnosis method based on local manifold learning of an autoencoded embedding," *IEEE Access*, vol. 9, pp. 30154–30168, 2021.

[108] T. Nakazawa and D. V. Kulkarni, "Wafer map defect pattern classification and image retrieval using convolutional neural network," *IEEE Transactions on Semiconductor Manufacturing*, vol. 31, no. 2, pp. 309–314, 2018.

[109] R. Sabir, D. Rosato, S. Hartmann and C. Guehmann, "LSTM based bearing fault diagnosis of electrical machines using motor current signal," in *Proceedings – 18th IEEE International Conference on Machine Learning and Applications, ICMLA 2019*, 2019, pp. 613–618.

[110] W. Tan, Y. Sun, D. Qiu, Y. An and P. Ren, "Rolling bearing fault diagnosis based on single gated unite recurrent neural networks," *Journal of Physics: Conference Series*, vol. 1601, no. 4, 2020.

[111] Y. Ding, L. Ma, J. Ma, C. Wang and C. Lu, "A generative adversarial network-based intelligent fault diagnosis method for rotating machinery under small sample size conditions," *IEEE Access*, vol. 7, pp. 149736–149749, 2019.

[112] J. Suwatthikul and S. Sornmuang, "Fault detection and diagnosis of a motor bearing shield," in *Proceedings of the 6th IEEE International Conference on Intelligent Data Acquisition and Advanced Computing Systems*, 2011, pp. 332–335.

[113] H. Abu-Rub, S. Moin Ahmed, A. Iqbal, M. Rahimian and H. A. Toliyat, "Incipient bearing fault diagnostics for inverter fed induction motor drive using ANFIS," in *The XIX International Conference on Electrical Machines – ICEM* 2010, 2010, pp. 1–5.

[114] M. Saeidi, J. Zarei, H. Hassani and A. Zamani, "Bearing fault detection via Park's vector approach based on ANFIS," in *2014 International Conference on Mechatronics and Control (ICMC)*, 2014, pp. 2436–2441

[115] J. Zarei, H. Hassani, Z. Wei and H. R. Karimi, "Broken rotor bars detection via Park's vector approach based on ANFIS," in *2014 IEEE 23rd*

International Symposium on Industrial Electronics (ISIE), 2014, pp. 2422–2426.

[116] Y. Lei, Z. He, Y. Zi and Q. Hu, "Fault diagnosis of rotating machinery based on multiple ANFIS combination with GAs," *Mechanical Systems and Signal Processing*, vol. 21, pp. 2280–2294, 2007.

[117] X. Lou and K. A. Loparo, "Bearing fault diagnosis based on wavelet transform and fuzzy inference," *Mechanical Systems and Signal Processing*, vol. 18, pp. 1077–1095, 2004.

[118] V. T. Tran, B. S. Yang, M. S. Oh and A. C. C. Tan, "Fault diagnosis of induction motor based on decision trees and adaptive neuro-fuzzy inference," *Expert Systems with Applications*, vol. 36, no. 2 Part 1, pp. 1840–1849, 2009.

[119] Q. T. Tran, S. D. Nguyen and T. Seo, "Algorithm for estimating online bearing fault upon the ability to extract meaningful information from big data of intelligent structures," *IEEE Transactions on Industrial Electronics*, vol. 66, no. 5, pp. 3804–3813, 2019.

[120] J. Sun, C. Yan and J. Wen, "Intelligent bearing fault diagnosis method combining compressed data acquisition and deep learning," *IEEE Transactions on Instrumentation and Measurement*, vol. 67, no. 1, pp. 185–195, 2018.

[121] J. J. Saucedo-Dorantes, A. Y. Jaen-Cuellar, M. Delgado-Prieto, R. J. Romero-Troncoso and R. A Osornio-Rios, "Condition monitoring strategy based on an optimized selection of high-dimensional set of hybrid features to diagnose and detect multiple and combined faults in an induction motor," *Measurement*, vol. 178, p. 109404, 2021.

[122] M. Ganchev, B. Kubicek and H. Kappeler, "Rotor temperature monitoring system," in *The XIX International Conference on Electrical Machines*, 2010, pp. 1–5.

[123] S. Stipetic, M. Kovacic, Z. Hanic and M. Vrazic, "Measurement of excitation winding temperature on synchronous generator in rotation using infrared thermography," *IEEE Transactions on Industrial Electronics*, vol. 59, no. 5, pp. 2288–2298, 2012.

[124] C. Mejuto, M. Mueller, M. Shanel, A. Mebarki, M. Reekie and D. Staton, "Improved synchronous machine thermal modelling," in *8th International Conference on Electrical Machines*, 2008, pp. 1–6.

[125] W. Kirchgässner, O. Wallscheid and J. Böcker, "Data-driven permanent magnet temperature estimation in synchronous motors with supervised machine learning," *IEEE Transactions on Energy Conversion*, vol. 36, no. 3, pp. 2059–2067, 2021.

[126] A. Choudhary, D. Goyal and S. S. Letha, "Infrared thermography-based fault diagnosis of induction motor bearings using machine learning," *IEEE Sensors Journal*, vol. 21, no. 2, pp. 1727–1734, 2021.

[127] W. Xu, Y. Wan, T. Zuo and X. Sha, "Transfer learning based data feature transfer for fault diagnosis," *IEEE Access*, vol. 8, pp. 76120–76129, 2020.

[128] T. Boukra, A. Lebaroud and G. Clerc, "Statistical and neural-network approaches for the classification of induction machine faults using the

ambiguity plane representation," *IEEE Transactions on Industrial Electronics*, vol. 60, no. 9, pp. 4034–4042, 2013.

[129] A. Abid, M. T. Khan and M. S. Khan, "Multidomain features-based GA optimized artificial immune system for bearing fault detection," *IEEE Transactions on Systems, Man, and Cybernetics: Systems*, vol. 50, no. 1, pp. 348–359, 2020.

[130] B. Xue, M. Zhang and W. N. Browne, "Particle swarm optimization for feature selection in classification: a multi-objective approach," *IEEE Transactions on Cybernetics*, vol. 43, no. 6, pp. 1656–1671, 2013.

[131] X. Z. Gao, S. J. Ovaska and X. Wang, "Genetic algorithms-based detector generation in negative selection algorithm," in *Proc. IEEE Mountain Workshop Adapt. Learn. Syst.*, Logan, UT, USA, 2006, pp. 133–137.

[132] X. Z. Gao, S. J. Ovaska, X. Wang and M.-Y. Chow, "Multi-level optimization of negative selection algorithm detectors with application in motor fault detection," *Intelligent Automation & Soft Computing*, vol. 16, no. 3, pp. 353–375, 2010.

[133] X. Z. Gao, H. Xu, X. Wang and K. Zenger, "A study of negative selection algorithm-based motor fault detection and diagnosis," *International Journal of Innovative Computing, Information and Control*, vol. 9, no. 2, pp. 875–901, 2013.

[134] A. Soualhi, G. Clerc and H. Razik, "Detection and diagnosis of faults in induction motor using an improved artificial ant clustering technique," *IEEE Transactions on Industrial Electronics*, vol. 60, no. 9, pp. 4053–4062, 2013.

[135] Z. Zhang, Z. Zhao and X. Li, *et al.*, "Faster multiscale dictionary learning method with adaptive parameter estimation for fault diagnosis of traction motor bearings," *IEEE Transactions on Instrumentation and Measurement*, vol. 70, 2021.

[136] S. B. Jiang, P. K. Wong and Y. C. Liang, "A fault diagnostic method for induction motors based on feature incremental broad learning and singular value decomposition," *IEEE Access*, vol. 7, pp. 157796–157806, 2019.

[137] C. Gong, Y. Hu, G. Chen, H. Wen, Z. Wang and K. Ni, "A DC-bus capacitor discharge strategy for PMSM drive system with large inertia and small system safe current in EVs," *IEEE Transactions on Industrial Informatics*, vol. 15, no. 8, pp. 4709–4718, 2019.

[138] H. Cherif, A. Benakcha and I. Laib, *et al.*, "Early detection and localization of stator inter-turn faults based on discrete wavelet energy ratio and neural networks in induction motor," *Energy*, vol. 212, p. 118684, 2020.

[139] P. C. M. Lamim Filho, R. Pederiva and J. N. Brito, "Detection of stator winding faults in induction machines using flux and vibration analysis," *Mechanical Systems and Signal Processing*, vol. 42, no. 1–2, pp. 377–387, 2014.

[140] M. Riera-Guasp, J. A. Antonino-Daviu and G. Capolino, "Advances in electrical machine, power electronic, and drive condition monitoring and fault detection: state of the art," *IEEE Transactions on Industrial Electronics*, vol. 62, no. 3, pp. 1746–1759, 2015.

[141] A. Sarikhani and O. A. Mohammed, "Inter-turn fault detection in PM synchronous machines by physics-based back electromotive force estimation," *IEEE Transactions on Industrial Electronics*, vol. 60, no. 8, pp. 3472–3484, 2013.

[142] L. Chen, J. Wang and Z. Sun, "Electromagnetic-thermal coupled modelling and analysis of inter-turn short-circuit faults of a permanent magnet alternator," *The Journal of Engineering*, vol. 2019, no. 17, pp. 4426–4431, 2019.

[143] J. Seshadrinath, B. Singh and B. K. Panigrahi, "Vibration analysis based interturn fault diagnosis in induction machines," *IEEE Transactions on Industrial Informatics*, vol. 10, no. 1, pp. 340–350, 2014.

[144] M. Singh and A. G. Shaik, "Incipient fault detection in stator windings of an induction motor using Stockwell transform and SVM," *IEEE Transactions on Instrumentation and Measurement*, vol. 69, no. 12, pp. 9496–9504, 2020.

[145] L. S. Maraaba, A. S. Milhem, I. A. Nemer, H. Al-Duwaish and M. A. Abido, "Convolutional neural network-based inter-turn fault diagnosis in LSPMSMs," *IEEE Access*, vol. 8, pp. 81960–81970, 2020.

[146] C. Kral, A. Haumer and S. B. Lee, "A practical thermal model for the estimation of permanent magnet and stator winding temperatures," *IEEE Transactions on Power Electronics*, vol. 29, no. 1, pp. 455–464, 2014.

[147] D. Reigosa, D. Fernandez, T. Tanimoto, T. Kato and F. Briz, "Comparative analysis of BEMF and pulsating high-frequency current injection methods for PM temperature estimation in PMSMs," *IEEE Transactions on Power Electronics*, vol. 32, no. 5, pp. 3691–3699, 2017.

[148] G. Feng, C. Lai, K. L. V. Iyer and N. C. Kar, "Improved high-frequency voltage injection based permanent magnet temperature estimation for PMSM condition monitoring for EV applications," *IEEE Transactions on Vehicular Technology*, vol. 67, no. 1, pp. 216–225, 2018.

[149] W. Kirchgässner, O. Wallscheid and J. Böcker, "Estimating electric motor temperatures with deep residual machine learning," *IEEE Transactions on Power Electronics*, vol. 36, no. 7, pp. 7480–7488, 2021.

[150] K.-H. Chao, L.-Y. Chang and F.-Q. Xu, "Three-level T-type inverter fault diagnosis and tolerant control using single-phase line voltage," *IEEE Access*, vol. 8, pp. 44075–44086, 2020.

[151] M. S. Haque, S. Choi and J. Baek, "Auxiliary particle filtering-based estimation of remaining useful life of IGBT," *IEEE Transactions on Industrial Electronics*, vol. 65, no. 3, pp. 2693–2703, 2018.

[152] Z. Rao, M. Huang and X. Zha, "IGBT remaining useful life prediction based on particle filter with fusing precursor," *IEEE Access*, vol. 8, pp. 154281–154289, 2020.

[153] S. H. Ali, M. Heydarzadeh, S. Dusmez, X. Li, A. S. Kamath and B. Akin, "Lifetime estimation of discrete IGBT devices based on Gaussian process," *IEEE Transactions on Industry Applications*, vol. 54, no. 1, pp. 395–403, 2018.

[154] A. Alghassi, S. Perinpanayagam and M. Samie, "Stochastic RUL calculation enhanced with TDNN-based IGBT failure modeling," *IEEE Transactions on Reliability*, vol. 65, no. 2, pp. 558–573, 2016.

[155] G. Sonnenfeld, K. Goebel and J. R. Celaya, "An agile accelerated aging, characterization and scenario simulation system for gate controlled power transistors," in *2008 IEEE AUTOTESTCON*, Salt Lake City, UT, USA, 2008.

[156] T. Wang, J. Qi, H. Xu, Y. Wang, L. Liu and D. Gao, "Fault diagnosis method based on FFT-RPCA-SVM for cascaded-multilevel inverter," *ISA Transactions*, vol. 60, pp. 156–163, 2016.

[157] S. S. Moosavi, A. Kazemi and H. Akbari, "A comparison of various open-circuit fault detection methods in the IGBT-based DC/AC inverter used in electric vehicle," *Engineering Failure Analysis*, vol. 96, pp. 223–235, 2019.

[158] Y. Cheng, W. Dong, F. Gao and G. Xin, "Open-circuit fault diagnosis of traction inverter based on compressed sensing theory," *Chinese Journal of Electrical Engineering*, vol. 6, no. 1, pp. 52–60, 2020.

[159] D. Chen, Y. Ye and R. Hua, "Fault diagnosis of NPC inverter based on multi-layer SVM," *Communications in Computer and Information Science*. Berlin, Heidelberg: Springer, pp. 611–621, 2012.

[160] H. Hu, F. Feng and T. Wang, "Open-circuit fault diagnosis of NPC inverter IGBT based on independent component analysis and neural network," *Energy Reports*, vol. 6, pp. 134–143, 2020.

[161] X. Wan, H. Hu, Y. Yu, L. Kang and F. Zeng, "Open-circuit fault diagnosis for grid-connected NPC inverter based on independent component analysis and neural network," *TELKOMNIKA (Telecommunication Computing Electronics and Control)*, vol. 15, no. 1, p. 36, 2017.

[162] D. Chen, Y. Ye and R. Hua, "Fault diagnosis system for NPC inverter based on multi-layer principal component neural network," *Journal of Computers*, vol. 8, no. 6, 2013.

[163] J. Liu, J. Wang, W. Yu, Z. Wang and G. Zhong, "Open-circuit fault diagnosis of traction inverter based on improved convolutional neural network," *Journal of Physics: Conference Series*, vol. 1633, p. 012099, 2020.

[164] Z. Ji and W. Liu, "Open-circuit fault detection for three-phase inverter based on backpropagation neural network," *Neural Computing and Applications*, vol. 31, no. 9, pp. 4665–4674, 2018.

[165] B. D. E. Cherif, A. Bendiabdellah and M. Tabbakh, "An automatic diagnosis of an inverter IGBT open-circuit fault based on HHT-ANN," *Electric Power Components and Systems*, vol. 48, no. 6–7, pp. 589–602, 2020.

[166] R. B. Dhumale and S. D. Lokhande, "Neural network fault diagnosis of voltage source inverter under variable load conditions at different frequencies," *Measurement*, vol. 91, pp. 565–575, 2016.

[167] W. Gong, H. Chen, Z. Zhang, M. Zhang and H. Gao, "A data-driven-based fault diagnosis approach for electrical power DC-DC inverter by using modified convolutional neural network with global average pooling and 2-D feature image," *IEEE Access*, vol. 8, pp. 73677–73697, 2020.

[168] R. K. Singleton, E. G. Strangas and S. Aviyente, "Extended Kalman filtering for remaining-useful-life estimation of bearings," *IEEE Transactions on Industrial Electronics*, vol. 62, no. 3, pp. 1781–1790, 2015.

[169] Y. Qian and R. Yan, "Remaining useful life prediction of rolling bearings using an enhanced particle filter," *IEEE Transactions on Instrumentation and Measurement*, vol. 64, no. 10, pp. 2696–2707, 2015.

[170] C. Zhang, P. Lim, A. K. Qin and K. C. Tan, "Multiobjective deep belief networks ensemble for remaining useful life estimation in prognostics," *IEEE Transactions on Neural Networks and Learning Systems*, vol. 28, no. 10, pp. 2306–2318, 2017.

[171] L. Ren, Y. Sun, H. Wang and L. Zhang, "Prediction of bearing remaining useful life with deep convolution neural network," *IEEE Access*, vol. 6, pp. 13041–13049, 2018.

[172] R. Liu, B. Yang and A. G. Hauptmann, "Simultaneous bearing fault recognition and remaining useful life prediction using joint-loss convolutional neural network," *IEEE Transactions on Industrial Informatics*, vol. 16, no. 1, pp. 87–96, 2020.

[173] J. Zhu, N. Chen and W. Peng, "Estimation of bearing remaining useful life based on multiscale convolutional neural network," *IEEE Transactions on Industrial Electronics*, vol. 66, no. 4, pp. 3208–3216, 2019.

[174] B. Wang, Y. Lei, N. Li and N. Li, "A Hybrid Prognostics Approach for Estimating Remaining Useful Life of Rolling Element Bearings," *IEEE Transactions on Reliability*, vol. 69, no. 1, pp. 401–412, 2020.

[175] Y. Yin, Z. Liu, M. Zuo, Z. Zhou, and J. Zhang, "A three-dimensional vibration data compression method for rolling bearing condition monitoring," *IEEE Transactions on Instrumentation and Measurement*, vol. 72, pp. 1–10, 2023.

[176] C. Lessmeier, J. K. Kimotho, D. Zimmer and W. Sextro, "Condition monitoring of bearing damage in electromechanical drive systems by using motor current signals of electric motors: a benchmark data set for data-driven classification," in *PHM Society European Conference*, vol. 3, no. 1, 2016.

[177] M. Zeng and Z. Chen, "SOSO boosting of the K-SVD denoising algorithm for enhancing fault-induced impulse responses of rolling element bearings," *IEEE Transactions on Industrial Electronics*, vol. 67, no. 2, pp. 1282–1292, 2020.

[178] Y. Qin, J. Zou, B. Tang and Y. Wang, "Transient feature extraction by the improved orthogonal matching pursuit and K-SVD algorithm with adaptive transient dictionary," *IEEE Transactions on Industrial Informatics*, vol. 16, no. 1, pp. 215–227, 2020.

[179] Z. Zhao, S. Wu, B. Qiao and S. Wang, "Enhanced sparse period-group lasso for bearing fault diagnosis," *IEEE Transactions on Industrial Electronics*, vol. 66, no. 3, pp. 2143–2153, 2019.

[180] F. Ben Abid, S. Zgarni and A. Braham, "Distinct bearing faults detection in induction motor by a hybrid optimized SWPT and aiNet-DAG SVM," *IEEE Transactions on Energy Conversion*, vol. 33, no. 4, pp. 1692–1699, 2018.

[181] M. Kang, J. Kim and J. Kim, "An FPGA-based multicore system for real-time bearing fault diagnosis using ultrasampling rate AE signals," *IEEE Transactions on Industrial Electronics*, vol. 62, no. 4, pp. 2319–2329, 2015.

[182] L. Wan, K. Gong, G. Zhang, X. Yuan, C. Li and X. Deng, "An efficient rolling bearing fault diagnosis method based on spark and improved random forest algorithm," *IEEE Access*, vol. 9, pp. 37866–37882, 2021.

[183] Z. Wang, Q. Zhang, J. Xiong, M. Xiao, G. Sun and J. He, "Fault diagnosis of a rolling bearing using wavelet packet denoising and random forests," *IEEE Sensors Journal*, vol. 17, no. 17, pp. 5581–5588, 2017.

[184] G. Xu, M. Liu, Z. Jiang, D. Söffker and W. Shen, "Bearing fault diagnosis method based on deep convolutional neural network and random forest ensemble learning," *Sensors*, vol. 19, no. 5, p. 1088, 2019.

[185] S. Guo, T. Yang, W. Gao, C. Zhang and Y. Zhang, "An intelligent fault diagnosis method for bearings with variable rotating speed based on Pythagorean spatial pyramid pooling CNN," *Sensors*, vol. 18, no. 11, p. 3857, 2018.

[186] Q. Jiang, F. Chang and B. Sheng, "Bearing fault classification based on convolutional neural network in noise environment," *IEEE Access*, vol. 7, pp. 69795–69807, 2019.

[187] L. Yuan, D. Lian, X. Kang, Y. Chen and K. Zhai, "Rolling bearing fault diagnosis based on convolutional neural network and support vector machine," *IEEE Access*, vol. 8, pp. 137395–137406, 2020.

[188] H. Li, J. Huang and S. Ji, "Bearing fault diagnosis with a feature fusion method based on an ensemble convolutional neural network and deep neural network," *Sensors*, vol. 19, no. 9, p. 2034, 2019.

[189] L. Hou, R. Jiang, Y. Tan and J. Zhang, "Input feature mappings-based deep residual networks for fault diagnosis of rolling element bearing with complicated dataset," *IEEE Access*, vol. 8, pp. 180967–180976, 2020.

[190] L. Chen, G. Xu, T. Tao and Q. Wu, "Deep residual network for identifying bearing fault location and fault severity concurrently," *IEEE Access*, vol. 8, pp. 168026–168035, 2020.

[191] L. Guo, Y. Lei, S. Xing, T. Yan and N. Li, "Deep convolutional transfer learning network: a new method for intelligent fault diagnosis of machines with unlabeled data," *IEEE Transactions on Industrial Electronics*, vol. 66, no. 9, pp. 7316–7325, 2019.

[192] Y. Wang, X. Ding, Q. Zeng, L. Wang and Y. Shao, "Intelligent rolling bearing fault diagnosis via vision ConvNet," *IEEE Sensors Journal*, vol. 21, no. 5, pp. 6600–6609, 2021.

[193] E. Tzeng, J. Hoffman, N. Zhang, K. Saenko and T. Darrell, "Deep domain confusion: maximizing for domain invariance," arXiv.org, 2014. [Online]. Available: https://arxiv.org/abs/1412.3474v1.

[194] Y. Ganin, E. Ustinova, and H. Ajakan, *et al.*, "Domain-adversarial training of neural networks," arXiv.org, 2016. [Online]. Available: https://arxiv.org/abs/1505.07818.

[195] S. J. Pan, I. W. Tsang, J. T. Kwok and Q. Yang, "Domain adaptation via transfer component analysis," *IEEE Transactions on Neural Networks*, vol. 22, no. 2, pp. 199–210, 2011.
[196] W. Lu, B. Liang, Y. Cheng, D. Meng, J. Yang and T. Zhang, "Deep model based domain adaptation for fault diagnosis," *IEEE Transactions on Industrial Electronics*, vol. 64, no. 3, pp. 2296–2305, 2017.
[197] H. Shao, M. Xia, G. Han, Y. Zhang and J. Wan, "Intelligent fault diagnosis of rotor-bearing system under varying working conditions with modified transfer convolutional neural network and thermal images," *IEEE Transactions on Industrial Informatics*, vol. 17, no. 5, pp. 3488–3496, 2021.
[198] A. Dosovitskiy, L. Beyer, A. Kolesnikov, *et al.*, "An image is worth 16x16 words: transformers for image recognition at scale," arXiv.org, 2021. [Online]. Available: https://arxiv.org/abs/2010.11929.
[199] A. Vaswani, N. Shazeer, N. Parmar, *et al.*, "Attention is all you need," arXiv.org, 2017. [Online]. Available: https://arxiv.org/abs/1706.03762v5. [Accessed 16 Jul 2021].
[200] J. L. Ba, J. R. Kiros and G. E. Hinton, "Layer normalization," 2016. [Online]. Available: https://arxiv.org/abs/1607.06450.
[201] D. Hendrycks and K. Gimpel, "Gaussian error linear units (GELUs)," arXiv.org, 2018. [Online]. Available: https://arxiv.org/abs/1606.08415v3.
[202] Z.-X. Hu, Y. Wang, M.-F. Ge and J. Liu, "Data-driven fault diagnosis method based on compressed sensing and improved multiscale network," *IEEE Transactions on Industrial Electronics*, vol. 67, no. 4, pp. 3216–3225, 2020.
[203] Y. Yin, Z. Liu, M. Zuo, Z. Zhou, and J. Zhang, "A three-dimensional vibration data compression method for rolling bearing condition monitoring," *IEEE Transactions on Instrumentation and Measurement*, vol. 72, p. 3506010, 2023.
[204] E. Tzeng, J. Hoffman, N. Zhang, K. Saenko and T. Darrell, "Deep domain confusion: Maximizing for domain invariance," arXiv:1412.3474, 2014. [Online].
[205] L. Guo, Y. Lei, S. Xing, T. Yan and N. Li, "Deep convolutional transfer learning network: a new method for intelligent fault diagnosis of machines with unlabeled data," *IEEE Transactions on Industrial Electronics*, vol. 66, no. 9, pp. 7316–7325, 2019.
[206] X. Wang, S. Lu, K. Chen, Q. Wang, and S. Zhang, "Bearing fault diagnosis of switched reluctance motor in electric vehicle powertrain via multisensor data fusion," *IEEE Transactions on Industrial Informatics*, vol. 18, no. 4, pp. 2452–2464, 2022.
[207] X. Jin, F. Cheng, Y. Peng, W. Qiao and L. Qu, "Drivetrain gearbox fault diagnosis: vibration- and current-based approaches," *IEEE Industry Applications Magazine*, vol. 24, no. 6, pp. 56–66, 2018.
[208] X. He, Q. Liu, W. Yu, C. K. Mechefske and X. Zhou, "A new autocorrelation-based strategy for multiple fault feature extraction from gearbox vibration signals," *Measurement*, vol. 171, p. 108738, 2021.

[209] Y. Lei, Z. Liu, X. Wu, N. Li, W. Chen and J. Lin, "Health condition identification of multi-stage planetary gearboxes using a mRVM-based method," *Mechanical Systems and Signal Processing*, vol. 60–61, pp. 289–300, 2015.

[210] Q. He, J. Zhao, G. Jiang and P. Xie, "An unsupervised multiview sparse filtering approach for current-based wind turbine gearbox fault diagnosis," *IEEE Transactions on Instrumentation and Measurement*, vol. 69, no. 8, pp. 5569–5578, 2020.

[211] H. Shi, J. Guo, X. Bai, L. Guo, Z. Liu and J. Sun, "Gearbox incipient fault detection based on deep recursive dynamic principal component analysis," *IEEE Access*, vol. 8, pp. 57646–57660, 2020.

[212] J. Zhang, J. S. Dhupia and C. J. Gajanayake, "Stator current analysis from electrical machines using resonance residual technique to detect faults in planetary gearboxes," *IEEE Transactions on Industrial Electronics*, vol. 62, no. 9, pp. 5709–5721, 2015.

[213] Y. Qin, S. Xiang, Y. Chai and H. Chen, "Macroscopic–microscopic attention in LSTM networks based on fusion features for gear remaining life prediction," *IEEE Transactions on Industrial Electronics*, vol. 67, no. 12, pp. 10865–10875, 2020.

[214] D.-F. Wang, Y. Guo, X. Wu, J. Na and G. Litak, "Planetary-gearbox fault classification by convolutional neural network and recurrence plot," *Applied Sciences*, vol. 10, no. 3, p. 932, 2020.

[215] C. Li, R.-V. Sanchez, G. Zurita, M. Cerrada, D. Cabrera and R. E. Vásquez, "Multimodal deep support vector classification with homologous features and its application to gearbox fault diagnosis," *Neurocomputing*, vol. 168, pp. 119–127, 2015.

[216] F. Chen, B. Tang and R. Chen, "A novel fault diagnosis model for gearbox based on wavelet support vector machine with immune genetic algorithm," *Measurement*, vol. 46, no. 1, pp. 220–232, 2013.

[217] D. Wang, "K-nearest neighbors based methods for identification of different gear crack levels under different motor speeds and loads: revisited," *Mechanical Systems and Signal Processing*, vol. 70–71, pp. 201–208, 2016.

[218] Y. Lei and M. J. Zuo, "Gear crack level identification based on weighted K nearest neighbor classification algorithm," *Mechanical Systems and Signal Processing*, vol. 23, no. 5, pp. 1535–1547, 2009.

[219] S. Zhu and B. Jiao, "Fault diagnosis of gearbox of wind turbine based on improved decision tree algorithm," in *Proceedings of the 2017 2nd International Conference on Control, Automation and Artificial Intelligence (CAAI 2017)*, 2017.

[220] J. Cheng and Y. Xiong, "Fault diagnosis of wind turbine gearbox based on improved QPSO algorithm," *Recent Advances in Electrical & Electronic Engineering (Formerly Recent Patents on Electrical & Electronic Engineering)*, vol. 12, no. 3, pp. 277–283, 2019.

[221] M. Azamfar, J. Singh, I. Bravo-Imaz and J. Lee, "Multisensor data fusion for gearbox fault diagnosis using 2-D convolutional neural network and

motor current signature analysis," *Mechanical Systems and Signal Processing*, vol. 144, p. 106861, 2020.

[222] G. Jiang, H. He, J. Yan and P. Xie, "Multiscale convolutional neural networks for fault diagnosis of wind turbine gearbox," *IEEE Transactions on Industrial Electronics*, vol. 66, no. 4, pp. 3196–3207, 2019.

[223] Z. Chen, C. Li and R.-V. Sanchez, "Gearbox fault identification and classification with convolutional neural networks," *Shock and Vibration*, vol. 2015, pp. 1–10, 2015.

[224] F. Cheng, Y. Peng, L. Qu and W. Qiao, "Current-based fault detection and identification for wind turbine drivetrain gearboxes," *IEEE Transactions on Industry Applications*, vol. 53, no. 2, pp. 878–887, 2017.

[225] F. Cheng, J. Wang, L. Qu and W. Qiao, "Rotor-current-based fault diagnosis for DFIG wind turbine drivetrain gearboxes using frequency analysis and a deep classifier," *IEEE Transactions on Industry Applications*, vol. 54, no. 2, pp. 1062–1071, 2018.

[226] D. Lu and W. Qiao, "Adaptive feature extraction and SVM classification for real-time fault diagnosis of drivetrain gearboxes," in *2013 IEEE Energy Conversion Congress and Exposition*, Denver, CO, USA, 2013.

[227] P. Shen, M. Ouyang, L. Lu, J. Li and X. Feng, "The co-estimation of state of charge, state of health, and state of function for lithium-ion batteries in electric vehicles," *IEEE Transactions on Vehicular Technology*, vol. 67, no. 1, pp. 92–103, 2018.

[228] F. Naseri, E. Farjah and T. Ghanbari, "An efficient regenerative braking system based on battery/supercapacitor for electric hybrid and plug-in hybrid electric vehicles with BLDC motor," *IEEE Transactions on Vehicular Technology*, vol. 66, no. 5, pp. 3724–3738, 2017.

[229] F. Naseri, E. Farjah, T. Ghanbari, Z. Kazemi, E. Schaltz and J. Schanen, "Online parameter estimation for supercapacitor state-of-energy and state-of-health determination in vehicular applications," *IEEE Transactions on Industrial Electronics*, vol. 67, no. 9, pp. 7963–7972, 2020.

[230] X. Hu, J. Jiang, D. Cao and B. Egardt, "Battery health prognosis for electric vehicles using sample entropy and sparse Bayesian predictive modeling," *IEEE Transactions on Industrial Electronics*, vol. 63, no. 4, pp. 2645–2656, 2016.

[231] J.-H. Lee, H.-S. Kim and I.-S. Lee, "Deep neural network based SOH monitoring of battery module," in *2019 IEEE Eurasia Conference on IOT, Communication and Engineering (ECICE)*, 2019.

[232] X. Feng, C. Weng and X. He, *et al.*, "Online state-of-health estimation for Li-ion battery using partial charging segment based on support vector machine," *IEEE Transactions on Vehicular Technology*, vol. 68, no. 9, pp. 8583–8592, 2019.

[233] W. Xiong, Y. Mo and C. Yan, "Online state-of-health estimation for second-use lithium-ion batteries based on weighted least squares support vector machine," *IEEE Access*, vol. 9, pp. 1870–1881, 2021.

[234] G. Crocioni, D. Pau, J. Delorme and G. Gruosso, "Li-ion batteries parameter estimation with tiny neural networks embedded on intelligent IoT microcontrollers," *IEEE Access*, vol. 8, pp. 122135–122146, 2020.

[235] Q. Huo, Z. Ma, X. Zhao, T. Zhang and Y. Zhang, "Bayesian network based state-of-health estimation for battery on electric vehicle application and its validation through real-world data," *IEEE Access*, vol. 9, pp. 11328–11341, 2021.

[236] L. Cai, J. Meng, D.-I. Stroe, J. Peng, G. Luo and R. Teodorescu, "Multiobjective optimization of data-driven model for lithium-ion battery SOH estimation with short-term feature," *IEEE Transactions on Power Electronics*, vol. 35, no. 11, pp. 11855–11864, 2020.

[237] D. Stroe and E. Schaltz, "Lithium-ion battery state-of-health estimation using the incremental capacity analysis technique," *IEEE Transactions on Industry Applications*, vol. 56, no. 1, pp. 678–685, 2020.

[238] E. Schaltz, D.-I. Stroe, K. Nørregaard, L. S. Ingvardsen and A. Christensen, "Incremental capacity analysis applied on electric vehicles for battery state-of-health estimation," *IEEE Transactions on Industry Applications*, vol. 57, no. 2, pp. 1810–1817, 2021.

[239] H. Chaoui and C. C. Ibe-Ekeocha, "State of charge and state of health estimation for lithium batteries using recurrent neural networks," *IEEE Transactions on Vehicular Technology*, vol. 66, no. 10, pp. 8773–8783, 2017.

[240] G. You, S. Park and D. Oh, "Diagnosis of electric vehicle batteries using recurrent neural networks," *IEEE Transactions on Industrial Electronics*, vol. 64, no. 6, pp. 4885–4893, 2017.

[241] J. Wu, X. Cui, H. Zhang and M. Lin, "Health prognosis with optimized feature selection for lithium-ion battery in electric vehicle applications," *IEEE Transactions on Power Electronics*, vol. 36, no. 11, pp. 12646–12655, 2021.

[242] H. Chaoui, C. Ibe-Ekeocha, A. Mejdoubi, A. Oukaour, H. Gualous and N. Omar, "State of charge estimation of $LiFePO_4$ batteries with temperature variations using neural networks," in *Proc. IEEE Int. Symp. Ind. Electron.*, 2016, pp. 286–291.

[243] L. Languang, X. Han, J. Li, J. Hua and M. Ouyang, "A review on the key issues for lithium-ion battery management in electric vehicles," *Journal of Power Sources*, vol. 226, pp. 272–288, 2013.

[244] Q. Wang, Y. He, J. Shen, X. Hu and Z. Ma, "State of charge-dependent polynomial equivalent circuit modeling for electrochemical impedance spectroscopy of lithium-ion batteries," *IEEE Transactions on Power Electronics*, vol. 33, no. 10, pp. 8449–8460, 2018.

[245] S. Skoog and S. David, "Parameterization of linear equivalent circuit models over wide temperature and SOC spans for automotive lithium-ion cells using electrochemical impedance spectroscopy," *Journal of Power Sources*, vol. 14, pp. 39–48, 2017.

[246] D. Xiao, G. Fang and S. Liu, *et al.*, "Reduced-coupling coestimation of SOC and SOH for lithium-ion batteries based on convex optimization,"

IEEE Transactions on Power Electronics, vol. 35, no. 11, pp. 12332–12346, 2020.

[247] N. Shateri, Z. Shi, D. J. Auger and A. Fotouhi, "Lithium-sulfur cell state of charge estimation using a classification technique," *IEEE Transactions on Vehicular Technology*, vol. 70, no. 1, pp. 212–224, 2021.

[248] R. Li, S. Xu and S. Li, *et al.*, "State of charge prediction algorithm of lithium-ion battery based on PSO-SVR cross validation," *IEEE Access*, vol. 8, pp. 10234–10242, 2020.

[249] J. C. Álvarez Antón, P. J. García Nieto, C. Blanco Viejo and J. A. Vilán Vilán, "Support vector machines used to estimate the battery state of charge," *IEEE Transactions on Power Electronics*, vol. 28, no. 12, pp. 5919–5926, 2013.

[250] L. Zhang, K. Li, D. Du, Y. Guo, M. Fei and Z. Yang, "A sparse learning machine for real-time SOC estimation of Li-ion batteries," *IEEE Access*, vol. 8, pp. 156165–156176, 2020.

[251] Y. Guo, Z. Yang, K. Liu, Y. Zhang and W. Feng, "A compact and optimized neural network approach for battery state-of-charge estimation of energy storage system," *Energy*, vol. 219, p. 119529, 2021.

[252] M. Wei, M. Ye, J. B. Li, Q. Wang and X. Xu, "State of charge estimation of lithium-ion batteries using LSTM and NARX neural networks," *IEEE Access*, vol. 8, pp. 189236–189245, 2020.

[253] B. Wang, C. Wang, Z. Wang, H. Xue and S. Ni, "Adaptive energy estimation for supercapacitor based on a real-time voltage state observer in electric vehicle applications," *IEEE Transactions on Power Electronics*, vol. 36, no. 7, pp. 7337–7341, 2021.

[254] P. Shrivastava, T. Kok Soon, M. Y. I. Bin Idris, S. Mekhilef and S. B. R. S. Adnan, "Combined state of charge and state of energy estimation of lithium-ion battery using dual forgetting factor-based adaptive extended Kalman filter for electric vehicle applications," *IEEE Transactions on Vehicular Technology*, vol. 70, no. 2, pp. 1200–1215, 2021.

[255] A. Nadeau, G. Sharma and T. Soyata, "State-of-charge estimation for supercapacitors: a Kalman filtering formulation," in *2014 IEEE International Conference on Acoustics, Speech and Signal Processing (ICASSP)*, 2014.

[256] A. El Mejdoubi, A. Oukaour, H. Chaoui, Y. Slamani, J. Sabor and H. Gualous, "Online supercapacitor diagnosis for electric vehicle applications," *IEEE Transactions on Vehicular Technology*, vol. 65, no. 6, pp. 4241–4252, 2016.

[257] X. Liu, J. Wu, C. Zhang and Z. Chen, "A method for state of energy estimation of lithium-ion batteries at dynamic currents and temperatures," *Journal of Power Sources*, vol. 270, pp. 151–157, 2014.

[258] Y. Zhang, R. Xiong, H. He and M. G. Pecht, "Lithium-ion battery remaining useful life prediction with box–cox transformation and Monte Carlo simulation," *IEEE Transactions on Industrial Electronics*, vol. 66, no. 2, pp. 1585–1597, 2019.

[259] L. Chen, J. Chen and H. Wang, *et al.*, "Remaining useful life prediction of battery using a novel indicator and framework with fractional grey model and unscented particle filter," *IEEE Transactions on Power Electronics*, vol. 35, no. 6, pp. 5850–5859, 2020.

[260] J. Wei, G. Dong and Z. Chen, "Remaining useful life prediction and state of health diagnosis for lithium-ion batteries using particle filter and support vector regression," *IEEE Transactions on Industrial Electronics*, vol. 65, no. 7, pp. 5634–5643, 2018.

[261] G. Dong, W. Han and Y. Wang, "Dynamic Bayesian network-based lithium-ion battery health prognosis for electric vehicles," *IEEE Transactions on Industrial Electronics*, vol. 68, no. 11, pp. 10949–10958, 2021.

[262] K. Park, Y. Choi, W. J. Choi, H. Ryu and H. Kim, "LSTM-based battery remaining useful life prediction with multi-channel charging profiles," *IEEE Access*, vol. 8, pp. 20786–20798, 2020.

[263] Y. Zhang, R. Xiong, H. He and M. G. Pecht, "Long short-term memory recurrent neural network for remaining useful life prediction of lithium-ion batteries," *IEEE Transactions on Vehicular Technology*, vol. 67, no. 7, pp. 5695–5705, 2018.

[264] K. Liu, Y. Shang, Q. Ouyang and W. D. Widanage, "A data-driven approach with uncertainty quantification for predicting future capacities and remaining useful life of lithium-ion battery," *IEEE Transactions on Industrial Electronics*, vol. 68, no. 4, pp. 3170–3180, 2021.

[265] B. Chinomona, C. Chung, L.-K. Chang, W.-C. Su and M.-C. Tsai, "Long short-term memory approach to estimate battery remaining useful life using partial data," *IEEE Access*, vol. 8, pp. 165419–165431, 2020.

[266] L. Yao, Z. Fang, Y. Xiao, J. Hou and Z. Fu, "An intelligent fault diagnosis method for lithium battery systems based on grid search support vector machine," *Energy*, vol. 214, p. 118866, 2021.

[267] R. Yang, R. Xiong, H. He and Z. Chen, "A fractional-order model-based battery external short circuit fault diagnosis approach for all-climate electric vehicles application," *Journal of Cleaner Production*, vol. 187, pp. 950–959, 2018.

[268] L. Yao, Y. Xiao, X. Gong, J. Hou and X. Chen, "A novel intelligent method for fault diagnosis of electric vehicle battery system based on wavelet neural network," *Journal of Power Sources*, vol. 453, p. 227870, 2020.

[269] R. Yang, R. Xiong, S. Ma and X. Lin, "Characterization of external short circuit faults in electric vehicle Li-ion battery packs and prediction using artificial neural networks," *Applied Energy*, vol. 260, p. 114253, 2020.

[270] J. Hong, Z. Wang and Y. Yao, "Fault prognosis of battery system based on accurate voltage abnormity prognosis using long short-term memory neural networks," *Applied Energy*, vol. 251, p. 113381, 2019.

[271] W. Gao, X. Li, M. Ma, Y. Fu, J. Jiang and C. Mi, "Case study of an electric vehicle battery thermal runaway and online internal short-circuit detection," *IEEE Transactions on Power Electronics*, vol. 36, no. 3, pp. 2452–2455, 2021.

[272] D. Li, Z. Zhang, P. Liu, Z. Wang and L. Zhang, "Battery fault diagnosis for electric vehicles based on voltage abnormality by combining the long short-term memory neural network and the equivalent circuit model," *IEEE Transactions on Power Electronics*, vol. 36, no. 2, pp. 1303–1315, 2021.

[273] Q. Xue, G. Li, Y. Zhang, S. Shen, Z. Chen and Y. Liu, "Fault diagnosis and abnormality detection of lithium-ion battery packs based on statistical distribution," *Journal of Power Sources*, vol. 482, p. 228964, 2021.

[274] O. Ojo, H. Lang, Y. Kim, X. Hu, B. Mu and X. Lin, "A neural network based method for thermal fault detection in lithium-ion batteries," *IEEE Transactions on Industrial Electronics*, vol. 68, no. 5, pp. 4068–4078, 2021.

[275] J. Zhu, T. Tan, L. Wu and H. Yuan, "RUL prediction of lithium-ion battery based on improved DGWO-ELM method in a random discharge rates environment," *IEEE Access*, vol. 7, pp. 125176–125187, 2019.

[276] J. S. L. Senanayaka, H. Van Khang and K. G. Robbersmyr, "Toward self-supervised feature learning for online diagnosis of multiple faults in electric powertrains," *IEEE Transactions on Industrial Informatics*, vol. 17, no. 6, pp. 3772–3781, 2021.

[277] H. Qiu, J. Lee, J. Lin and G. Yu, "Wavelet filter-based weak signature detection method and its application on rolling element bearing prognostics," *Journal of Sound and Vibration*, vol. 289, no. 4, pp. 1066–1090, 2006.

[278] S. W. Wade and R. B. Randall, "Rolling element bearing diagnostics using the Case Western Reserve University data: a benchmark study," *Mechanical Systems and Signal Processing*, vol. 64–65, pp. 100–131, 2015.

[279] Y.-S. Lai, W.-T. Lee, Y.-K. Lin, and J.-F. Tsai, "Integrated inverter/converter circuit and control technique of motor drives with dual-mode control for EV/HEV applications," *IEEE Transactions on Power Electronics*, vol. 29, no. 3, pp. 1358–1365, 2014.

[280] S. Suroso, D. T. Nugroho and T. Noguchi, "New dead-time compensation method of power inverter using carrier based sinusoidal pulse-width modulation," *International Journal of Electrical and Computer Engineering, IJECE*, vol. 8, no. 6, p. 4880, 2018, doi: 10.11591/ijece.v8i6.pp4880-4891.

[281] Y. Tadano, S. Urushibata, K. Ogura, A. Shigaki and M. Nomura, "A multilevel PWM strategy suitable for high-voltage motor direct drive systems in consideration of the adverse effect of a deadtime," *IEEJ Transactions on Industry Applications*, vol. 126, no. 1, pp. 1–9, 2006, doi: 10.1541/ieejias.126.1.

[282] J.-W. Choi and S.-K. Sul, "Inverter output voltage synthesis using novel dead time compensation," *IEEE Transactions on Power Electronics*, vol. 11, no. 2, pp. 221–227, 1996, doi: 10.1109/63.486169.

[283] S. Emad, R. Moustafa, and R. Hassan, *et al.*, "A comparison between different PWM methods of three phase Z-inverter for renewable energy applications," *Solid State Technology*, vol. 63, pp. 5462, 2020.

[284] Z. Wu, H. Zhang and K. Liu, *et al.*, "Dead-time compensation based on a modified multiple complex coefficient filter for permanent magnet synchronous machine drives," *IEEE Transactions on Power Electronics*, vol. 36, no. 11, pp. 12979–12989, 2021, doi: 10.1109/TPEL.2021.3078167.

[285] J. Zhou, K. Liu, J. Li, L. Li, W. Hu and R. Ding, "VSI nonlinearity compensation of a PMSM drive system using deadbeat prediction based current zero-crossing detection," *World Electric Vehicle Journal*, vol. 12, no. 1, p. 17, 2021, doi: 10.3390/wevj12010017.

[286] K. Yu and Z. Wang, "An online compensation method of VSI nonlinearity for dual three-phase PMSM drives using current injection," *IEEE Transactions on Power Electronics*, vol. 37, no. 4, pp. 3769–3774, 2022, doi: 10.1109/TPEL.2021.3127990.

[287] D.-H. Lee and J.-W. Ahn, "A simple and direct dead-time effect compensation scheme in PWM-VSI," *IEEE Transactions on Industry Applications*, vol. 50, no. 5, pp. 3017–3025, 2014, doi: 10.1109/TIA.2014.2303932.

[288] Z. Shen and D. Jiang, "Dead-time effect compensation method based on current ripple prediction for voltage-source inverters," *IEEE Transactions on Power Electronics*, vol. 34, no. 1, pp. 971–983, 2019, doi: 10.1109/TPEL.2018.2820727.

[289] A. R. Munoz and T. A. Lipo, "On-line dead-time compensation technique for open-loop PWM-VSI drives," *IEEE Transactions on Power Electronics*, vol. 14, no. 4, pp. 683–689, 1999, doi: 10.1109/63.774205.

[290] A. Lewicki, "Dead-time effect compensation based on additional phase current measurements," *IEEE Transactions on Industrial Electronics*, vol. 62, no. 7, pp. 4078–4085, 2015, doi: 10.1109/TIE.2015.2389756.

[291] J. Zhang and F. L. Luo, "An accurate approach of dead-time compensation for three-phase DC/AC inverter," in *2009 4th IEEE Conference on Industrial Electronics and Applications*, May 2009, pp. 2929–2934, doi: 10.1109/ICIEA.2009.5138745.

[292] Z. Zhang and L. Xu, "Dead-time compensation of inverters considering snubber and parasitic capacitance," *IEEE Transactions on Power Electronics*, vol. 29, no. 6, pp. 3179–3187, 2014, doi: 10.1109/TPEL.2013.2275551.

[293] Y. Park and S.-K. Sul, "Implementation schemes to compensate for inverter nonlinearity based on trapezoidal voltage," *IEEE Transactions on Industry Applications*, vol. 50, no. 2, pp. 1066–1073, 2014, doi: 10.1109/TIA.2013.2272431.

[294] D. E. Salt, D. Drury, D. Holliday, A. Griffo, P. Sangha and A. Dinu, "Compensation of inverter nonlinear distortion effects for signal-injection-based sensorless control," *IEEE Transactions on Industry Applications*, vol. 47, no. 5, pp. 2084–2092, 2011, doi: 10.1109/TIA.2011.2161430.

[295] G. Shen, W. Yao, B. Chen, K. Wang, K. Lee and Z. Lu, "Auto measurement of the inverter output voltage delay curve to compensate for inverter nonlinearity in sensorless motor drives," *IEEE Transactions on Power Electronics*, vol. 29, no. 10, pp. 5542–5553, 2014, doi: 10.1109/TPEL.2013.2293134.

[296] T. Mannen and H. Fujita, "Dead-time compensation method based on current ripple estimation," *IEEE Transactions on Power Electronics*, vol. 30, no. 7, pp. 4016–4024, 2015, doi: 10.1109/TPEL.2014.2352716.

[297] W. Dafang, Y. Bowen, Z. Cheng, Z. Chuanwei and Q. Ji, "A feedback-type phase voltage compensation strategy based on phase current reconstruction

for ACIM drives," *IEEE Transactions on Power Electronics*, vol. 29, pp. 5031–5043, 2014.

[298] Y.-K. Lin and Y.-S. Lai, "Dead-time elimination of PWM-controlled inverter/converter without separate power sources for current polarity detection circuit," *IEEE Transactions on Industrial Electronics*, vol. 56, no. 6, pp. 2121–2127, 2009, doi: 10.1109/TIE.2009.2014305.

[299] Y. Murai, T. Watanabe and H. Iwasaki, "Waveform distortion and correction circuit for PWM inverters with switching lag-times," *IEEE Transactions on Industry Applications*, vol. IA-23, no. 5, pp. 881–886, 1987, doi: 10.1109/TIA.1987.4504998.

[300] D. Leggate and R. J. Kerkman, "Pulse-based dead-time compensator for PWM voltage inverters," *IEEE Transactions on Industrial Electronics*, vol. 44, no. 2, pp. 191–197, 1997, doi: 10.1109/41.564157.

[301] J.-L. Lin, "A new approach of dead-time compensation for PWM voltage inverters," *IEEE Transactions on Circuits and Systems I: Fundamental Theory and Applications*, vol. 49, no. 4, pp. 476–483, 2002, doi: 10.1109/81.995662.

[302] D. B. Rathnayake, S. M. H. K. Samarasinghe, C. I. Medagedara and S. G. Abeyratne, "An enhanced pulse-based dead-time compensation technique for PWM-VSI drives," in *2014 9th International Conference on Industrial and Information Systems (ICIIS)*, Dec. 2014, pp. 1–5, doi: 10.1109/ICIINFS.2014.7036532.

[303] Y. Zhao, W. Qiao and L. Wu, "Dead-time effect and current regulation quality analysis for a sliding-mode position observer-based sensorless IPMSM drives," in *2012 IEEE Industry Applications Society Annual Meeting*, Oct. 2012, pp. 1–7, doi: 10.1109/IAS.2012.6374042.

[304] V. Blahník, Z. Peroutka, J. Žák and J. Talla, "Elimination of dead-time effect causing current distortion of single-phase power converters," in *2012 15th International Power Electronics and Motion Control Conference (EPE/PEMC)*, Sep. 2012, pp. DS1e.3-1–DS1e.3-6, doi:10.1109/EPEPEMC.2012.6397243.

[305] Ü. Büyükkeleş and A. Ersak, "Analysis and compensation of nonlinear effects in an inverter with power MOSFETs," in *2012 15th International Power Electronics and Motion Control Conference (EPE/PEMC)*, Sep. 2012, pp. DS2c.3-1–DS2c.3-6, doi: 10.1109/EPEPEMC.2012.6397287.

[306] G. Wang, R. Yang and D. Xu, "DSP-based control of sensorless IPMSM drives for wide-speed-range operation," *IEEE Transactions on Industrial Electronics*, vol. 60, no. 2, pp. 720–727, 2013, doi: 10.1109/TIE.2012.2205360.

[307] S. Jin and Y. Zhong, "A novel three-level SVPWM algorithm considering neutral-point control, narrow-pulse elimination and dead-time compensation," in *The 4th International Power Electronics and Motion Control Conference, 2004. IPEMC 2004*, vol. 2, Aug. 2004, pp. 688–693.

[308] C. Attaianese and G. Tomasso, "Predictive compensation of dead-time effects in VSI feeding induction motors," *IEEE Transactions on Industry Applications*, vol. 37, no. 3, pp. 856–863, 2001, doi: 10.1109/28.924768.

[309] G. Liu, D. Wang, Y. Jin, M. Wang and P. Zhang, "Current-detection-independent dead-time compensation method based on terminal voltage A/D conversion for PWM VSI," *IEEE Transactions on Industrial Electronics*, vol. 64, no. 10, pp. 7689–7699, 2017, doi: 10.1109/TIE.2017.2696480.

[310] L. Chen and F. Z. Peng, "Dead-time elimination for voltage source inverters," *IEEE Transactions on Power Electronics*, vol. 23, no. 2, pp. 574–580, 2008, doi: 10.1109/TPEL.2007.915766.

[311] X. Zhang, Y. Cheng, Z. Zhao and K. Yan, "Optimized model predictive control with dead-time voltage vector for PMSM drives," *IEEE Transactions on Power Electronics*, vol. 36, no. 3, pp. 3149–3158, 2021, doi: 10.1109/TPEL.2020.3012985.

[312] Z. Zhang, H. Lu, D. J. Costinett, F. Wang, L. M. Tolbert and B. J. Blalock, "Model-based dead time optimization for voltage-source converters utilizing silicon carbide semiconductors," *IEEE Transactions on Power Electronics*, vol. 32, no. 11, pp. 8833–8844, 2017, doi: 10.1109/TPEL.2016.2645578.

[313] N. Bedetti, S. Calligaro and R. Petrella, "Self-commissioning of inverter dead-time compensation by multiple linear regression based on a physical model," *IEEE Transactions on Industry Applications*, vol. 51, no. 5, pp. 3954–3964, 2015, doi: 10.1109/TIA.2015.2436882.

[314] K. Liu and Z. Q. Zhu, "Online estimation of the rotor flux linkage and voltage-source inverter nonlinearity in permanent magnet synchronous machine drives," *IEEE Transactions on Power Electronics*, vol. 29, no. 1, pp. 418–427, 2014, doi: 10.1109/TPEL.2013.2252024.

[315] S.-Y. Kim, W. Lee, M.-S. Rho and S.-Y. Park, "Effective dead-time compensation using a simple vectorial disturbance estimator in PMSM drives," *IEEE Transactions on Industrial Electronics*, vol. 57, no. 5, pp. 1609–1614, 2010, doi: 10.1109/TIE.2009.2033098.

[316] H.-S. Kim, H.-T. Moon and M.-J. Youn, "On-line dead-time compensation method using disturbance observer," *IEEE Transactions on Power Electronics*, vol. 18, no. 6, pp. 1336–1345, 2003, doi: 10.1109/TPEL.2003.818833.

[317] N. Urasaki, T. Senjyu, K. Uezato and T. Funabashi, "Adaptive dead-time compensation strategy for permanent magnet synchronous motor drive," *IEEE Transactions on Energy Conversion*, vol. 22, no. 2, pp. 271–280, 2007, doi: 10.1109/TEC.2006.875469.

[318] N. Urasaki, T. Senjyu, K. Uezato and T. Funabashi, "An adaptive dead-time compensation strategy for voltage source inverter fed motor drives," *IEEE Transactions on Power Electronics*, vol. 20, no. 5, pp. 1150–1160, 2005, doi: 10.1109/TPEL.2005.854046.

[319] M. A. Herran, J. R. Fischer, S. A. Gonzalez, M. G. Judewicz and D. O. Carrica, "Adaptive dead-time compensation for grid-connected PWM inverters of single-stage PV systems," *IEEE Transactions on Power Electronics*, vol. 28, no. 6, pp. 2816–2825, 2013, doi: 10.1109/TPEL.2012.2227811.

[320] S.-Y. Kim and S.-Y. Park, "Compensation of dead-time effects based on adaptive harmonic filtering in the vector-controlled AC motor drives," *IEEE Transactions on Industrial Electronics*, vol. 54, no. 3, pp. 1768–1777, 2007, doi: 10.1109/TIE.2007.895132.

[321] T. Qiu, X. Wen and F. Zhao, "Adaptive-linear-neuron-based dead-time effects compensation scheme for PMSM drives," *IEEE Transactions on Power Electronics*, vol. 31, no. 3, pp. 2530–2538, 2016, doi: 10.1109/TPEL.2015.2427914.

[322] S.-H. Hwang and J.-M. Kim, "Dead time compensation method for voltage-fed PWM inverter," *IEEE Transactions on Energy Conversion*, vol. 25, no. 1, pp. 1–10, 2010, doi: 10.1109/TEC.2009.2031811.

[323] Y. Yang, K. Zhou, H. Wang and F. Blaabjerg, "Harmonics mitigation of dead time effects in PWM converters using a repetitive controller," in *2015 IEEE Applied Power Electronics Conference and Exposition (APEC)*, Mar. 2015, pp. 1479–1486, doi: 10.1109/APEC.2015.7104543.

[324] B. A. Francis and W. M. Wonham, "The internal model principle for linear multivariable regulators," *Applied Mathematics and Optimization*, vol. 2, pp. 170–194, 1975.

[325] Z. Tang and B. Akin, "Suppression of dead-time distortion through revised repetitive controller in PMSM drives," *IEEE Transactions on Energy Conversion*, vol. 32, no. 3, pp. 918–930, 2017, doi: 10.1109/TEC.2017.2679701.

[326] F. Blaabjerg, J. K. Pedersen and P. Thoegersen, "Improved modulation techniques for PWM-VSI drives," *IEEE Transactions on Industrial Electronics*, vol. 44, no. 1, pp. 87–95, 1997, doi: 10.1109/41.557503.

[327] J. Zheng, Z. Wang, D. Wang, Y. Li and M. Li, "Review of fault diagnosis of PMSM drive system in electric vehicles," in *2017 36th Chinese Control Conference (CCC)*, Jul. 2017, pp. 7426–7432, doi: 10.23919/ChiCC.2017.8028529.

[328] Y. Wang, K. Liu, W. Hua, C. Zhang, Z. Wu and H. Zhang, "Analysis and detection of rotor eccentricity in permanent magnet synchronous machines based on linear hall sensors," *IEEE Transactions on Power Electronics*, vol. 37, no. 4, pp. 4719–4729, 2022, doi: 10.1109/TPEL.2021.3131576.

[329] M. A. Awadallah and M. M. Morcos, "Application of AI tools in fault diagnosis of electrical machines and drives-an overview," *IEEE Transactions on Energy Conversion*, vol. 18, no. 2, pp. 245–251, 2003, doi: 10.1109/TEC.2003.811739.

[330] C. Ma, J. Li, N. Zhang, F. Bu and Z. Yang, "Open-circuit radial stray magnetic flux density based non-invasive diagnosis for mixed eccentricity parameters of interior permanent magnet synchronous motors in electric vehicles," *IEEE Transactions on Industrial Electronics*, pp. 1–1, 2022, doi: 10.1109/TIE.2022.3165250.

[331] S. Nandi, H. A. Toliyat and X. Li, "Condition monitoring and fault diagnosis of electrical motors—a review," *IEEE Transactions on Energy*

Conversion, vol. 20, no. 4, pp. 719–729, 2005, doi: 10.1109/TEC.2005.847955.

[332] W. Lang, Y. Hu, C. Gong, X. Zhang, H. Xu and J. Deng, "Artificial intelligence-based technique for fault detection and diagnosis of EV motors: a review," *IEEE Transactions on Transportation Electrification*, vol. 8, no. 1, pp. 384–406, 2022, doi: 10.1109/TTE.2021.3110318.

[333] Y. Han, G. Huang, S. Song, L. Yang, H. Wang and Y. Wang, "Dynamic neural networks: a survey," *IEEE Transactions on Pattern Analysis and Machine Intelligence*, vol. 44, no. 11, pp. 7436–7456, 2022, doi: 10.1109/TPAMI.2021.3117837.

[334] X. Zhang, Y. Hu, A. Yin, J. Deng, H. Xu and J. Si, "Inferable deep distilled attention network for diagnosing multiple motor bearing faults," *IEEE Transactions on Transportation Electrification*, vol. 9, no. 2, pp. 2207–2216, 2023, doi: 10.1109/TTE.2022.3211203.

[335] X. Zhang, Y. Hu, J. Deng, H. Xu and H. Wen, "Feature engineering and artificial intelligence-supported approaches used for electric powertrain fault diagnosis: a review," *IEEE Access*, vol. 10, pp. 29069–29088, 2022, doi: 10.1109/ACCESS.2022.3157820.

[336] X. Zhang, Y. Hu, C. Gong, J. Deng and G. Wang, "Artificial intelligence technique based EV powertrain condition monitoring and fault diagnosis: a review," *IEEE Sensors Journal*, doi: 10.1109/JSEN.2023.3285531.

Index

Printed in the USA
CPSIA information can be obtained
at www.ICGtesting.com
JSHW012020200524
63488JS00003B/121

9 781839 537622